Statistics for Problem Solving

Statistics for Problem Solving

C. *Mitchell* Dayton

Chauncey

ASSOCIATE PROFESSOR OF EDUCATION
UNIVERSITY OF MARYLAND

Clayton L. Stunkard

PROFESSOR OF EDUCATION
UNIVERSITY OF MARYLAND

McGraw-Hill Book Company

NEW YORK   ST. LOUIS   SAN FRANCISCO   DÜSSELDORF   JOHANNESBURG
KUALA LUMPUR   LONDON   MEXICO   MONTREAL   NEW DELHI
PANAMA   RIO DE JANEIRO   SINGAPORE   SYDNEY   TORONTO

Statistics for Problem Solving

Library of Congress Catalog Card Number 79-140955

07-016182-8

1 2 3 4 5 6 7 8 9 0 M A M M 7 9 8 7 6 5 4 3 2 1 0

This book was set in Baskerville by The Maple Press
Company, and printed on permanent paper and bound
by The Maple Press Company. The drawings were
done by John Cordes, J. & R. Technical Services, Inc.
The editors were William J. Willey and Andrea
Stryker-Rodda. John F. Harte supervised production.

Contents

Preface

This volume is intended to serve as the primary textbook for a one-semester course in introductory applied statistics for students in education, educational psychology, and related behavioral science fields. Selection of content and level of presentation of material was designed to permit the development of a meaningful basis for continued study of applied statistics at the intermediate and advanced levels, while enabling the beginning student to become an informed consumer of research that utilizes the methodology of statistical inference. In view of the varied backgrounds and interests of beginning statistics students, the mathematical level of the treatment is such that only a good grasp of elementary algebra is required.

As the title of this volume implies, emphasis is placed on the utilization of statistical procedures in the analysis of research data. While many textbooks treat the fields of descriptive and inferential statistics separately, our approach has been to integrate their treatment and to focus on the use of statistics as the basis for a problem-solving strategy. From a didactic point of view, there are advantages to the early introduction of statistical inference since the logic of inference is ordinarily a difficult concept for beginning students; early treatment of this logic can permit adequate time during the course for repeated explanation and illustration, which increases the likelihood that the student will acquire a clear understanding and appreciation for statistical inference by the end of the course.

The organization of the textbook is in terms of a particular conceptualization of a framework for the field of applied statistics. This organization moves from simple to complex in four dimensions: purpose of the statistical analysis (description or inference); type of data to be analyzed (categorical, rank order, or measured); number of groups of data, or individuals, to be analyzed (one group, or two or more groups); and number of quantified variables to be analyzed (one, two, or more than two). This organization is developed in some detail in Chapter 1 and the major motivation for the selection of this organizational scheme was that these four dimensions provide a convenient framework for the choice of appropriate statistical tools in actual research situations.

In this age of high-speed electronic computers, authors of statistics textbooks (as well as classroom instructors) must carefully consider their approach to computational procedures for the statistical tools which they introduce to students. Procedures which are convenient for hand computation are ordinarily not best suited to the use of calculating machines and, similarly, calculating machine routines are often not the most efficient when programming for an electronic computer. We have attempted to select computational procedures which provide a high degree of insight into the nature of the statistical quantities being calculated; in some instances, the classroom instructor may wish to supplement our presentations with formulas in a form more amenable to hand, calculating machine, or computer utilization when this approach is better adapted to his own classroom situation. In our own beginning-level quantitative methods sequence, students are introduced to the use of canned computer programs at an early point and many of the end-of-chapter exercises are assigned to be completed as computer problems. However, calculations for the solutions to these end-of-chapter exercises (which are given in the back of the textbook) have been carried out using the formulas as presented in the textual material and may differ slightly from computer solutions.

We have included five Technical Appendices which treat selected topics which may be of interest in some courses. We assume that most instructors would wish to include Technical Appendix A (Units of Measurement and Significant Digits) and Technical Appendix C (The t Test for the Two-sample Case) at appropriate points in the course; the other Technical Appendices often may be of interest to specific students in a course, although not as a general requirement. Also, Chapter 8, which treats procedures applicable to rank-order data, may be omitted in a one-semester course without loss of continuity.

Earlier versions of this textbook (in mimeographed form) have been used for several years in our elementary statistics courses at the University of Maryland, and we wish to express our gratitude to the many students whose comments and forebearance have resulted in progressive improvements to the content and expression of this volume. Also, several faculty colleagues have utilized these earlier versions and their considered criticisms have been very valuable during our final revisions. Finally, we thank Dr. E. S. Pearson who kindly allowed us to reproduce percentage points of the F_{max} distributions for our tables in Appendix F.

C. Mitchell Dayton
Clayton L. Stunkard

Statistics for Problem Solving

One

introduction

1.1 General types of statistical operations

The purpose of applied statistics is to provide methods for solving practical problems. These problems concern the nature of the characteristics of some defined group of objects, organisms, events, or measurements. In the behavioral sciences, applied statistics is typically utilized to attack problems concerning living organisms, specifically, human beings. Some problems that may be approached by the use of statistical methods deal simply with isolated characteristics of people. For example, in attempting to justify a request for salary increases for high school principals in a certain city, a researcher may determine an average present salary for these principals. Characteristics such as average salary of principals, population of a city, and percentage of illiteracy in a state are often studied by statistical methods. Many people automatically associate the word statistics with compilations or summaries of data in the form of tables or graphs (of which a world almanac is an excellent example). Whereas these types of statistical procedures are widely known and of great practical importance, this book will emphasize statistical procedures that go beyond simple summaries and compilations of data. Statistical procedures in a broader sense may be used to attack problems concerning the relationships among the characteristics of organisms, objects, events, etc. Typical problems involving such relationships include: determining the characteristics of parents and home environment associated with juvenile delinquency; comparing two methods of classroom instruction to determine which, if either, results in more efficient learning by students; testing deductions from two opposing theories of learning concerning the behavior of students studying programmed instruction; determining how well scores on a test of creativity can be predicted from a knowledge of students' scores on an intelligence test.

Whether the researcher is studying the isolated characteristics of organisms or the relationships among characteristics of organisms, there are two general types of situations in which he can apply statistical methods. In the first situation, a problem exists concerning some specific available group of objects or organisms. "Available group" means that the researcher is able to perform measurements and other operations directly on *each* member of the group. Furthermore, the available group of objects or organisms must be the only objects or organisms that are of immediate interest to the researcher. The available group in this situation is referred to as the *population* of interest; and the statistical techniques that are appropriate are called *descriptive statistics*. Some practical situations that call for the application of descriptive statistical methods are: determining an average age for a specific classroom of students at a given time; determining how variable family incomes are for a specific community during one work year; discovering whether or not the high-ability students in a given school are also above average in sociability.

For the second general type of situation, only a subgroup, or *sample*, from the population is considered, not the entire population of interest. The statistical techniques appropriate for this situation are referred to as *inferential statistics*. A necessary condition for the use of inferential statistics is that the available sample be selected from the total population in a random manner. Because a precise definition of a random sample will be presented later, it is sufficient for the moment to regard a random sample as one in which the objects or organisms are selected by chance from the population. The use of a sample rather than the entire population can be a matter of either choice or necessity. From a practical point of view, time and cost factors can often be greatly reduced by use of random samples. Information from a random sample is often as accurate as practical purposes demand. Also, in some situations, the use of complete populations would render the research meaningless or, at least, very difficult to interpret. For example, it may be of interest to compare two procedures for teaching a certain motor skill. Call these two procedures method *A* and method *B*. If the whole population were first taught by method *A*, then no reasonable evaluation of method *B* could be made on this same population since the prior learning under method *A* could not be erased. However, each of two randomly formed groups could be given instruction by one of the methods and then compared to see which method was more effective for teaching the motor skill.

It should be noted that the techniques of both descriptive and inferential statistics are basically concerned with making statements about characteristics of the population of interest. Descriptive procedures utilize *each* member of the population in order to *describe* these characteristics; inferential procedures use only a sample from the population in order to *infer* these characteristics. To clarify this point, suppose that a researcher is interested in the social behavior of high school students who are entering a large university as freshmen. Specifically, suppose that this researcher would like to know what percentage of entering freshmen held leadership positions (e.g., class officer, president of a school club) while in high school. The population of interest consists of all entering freshmen at the university in the given year. The problem of determining what percentage of entering freshmen had leadership experience in high school could, in theory, be attacked by either descriptive or inferential procedures. If the researcher planned his study well in advance of the fall registration period at the university, he could possibly arrange to have *each* freshman fill out a questionnaire about his high school leadership experience. The number of freshmen claiming to have held leadership positions could then be found and the required percentage could be computed. (Note that this procedure would find the percentage of students who *claimed* that they held leadership positions. To verify these student reports would require contacting the high schools that the students attended. In many studies, the researcher may be satisfied with the student's own report.) Now, suppose that the researcher formulated his study too late to administer the questionnaire during registration. To contact each freshman would be rather expensive, and many freshmen might fail to return a questionnaire mailed to them. To offset these problems, the researcher could select a random sample from all the freshmen and utilize inferential techniques. He could, for example, select 200 freshmen randomly and then interview these students in their dormitories to collect the data on their high school leadership experience. If 22 percent of the freshmen *in the sample* said that they had held leadership positions, this percentage would be only an estimate of the actual percentage in the total population of freshmen. The population percentage could be either greater or smaller than 22 percent even though 22 percent report leadership experience in the sample. Inferential statistics provides procedures for making decisions concerning the nature of the population in cases of this sort, but each decision may entail some known amount of risk of being incorrect.

In summary, *both description and inferential statistical procedures have as their goal statements concerning populations of interest.* The methods of descriptive statistics are employed whenever entire populations of interest are studied by the researcher; the methods of inferential statistics are utilized whenever decisions are based on samples from populations of interest.

1.2 Measurement processes

All statistical procedures deal with numbers obtained from a process of measurement. The term "measurement" is interpreted in a broader sense in the behavioral sciences than it is in the physical sciences. Essentially, measurement involves a systematic procedure for assigning numbers to objects, organisms, or events.† A "systematic procedure" of this sort is referred to as a *measurement rule*. For example, the statement "assign the number 1 to males and the number 2 to females" is a measurement rule, since it specifies a systematic procedure for assigning numbers to organisms.‡ Part of any measurement rule must indicate the specific aspect, or characteristic, of the objects, organisms, or events to which the rule applies. The measurement rule presented above as an example indicates that the sex of the organism is the characteristic to which the rule applies. In order for measurement to be meaningful, the characteristic that is specified must take on two or more different values or forms. If such is not the case (i.e., if the characteristic comes in just one form), then measurement involves assigning exactly the same number to each object, organism, or event. A characteristic appearing in two or more forms is referred to as a *variable*. Thus, sex of mammals is a variable, since it occurs in two forms, male and female. A variable that occurs in exactly two forms is said to be *dichotomous;* one that occurs in exactly three forms is said to be *trichotomous*. In general, if a variable takes on only a limited number of forms or values within a specified range of the variable, it is referred to as a *discrete* variable. Sex, college course enrollment, and number of children in a family are examples of discrete variables, since each characteristic takes on only a limited number of values. Thus, a family may contain 3 children or 4 children, but $3\frac{1}{8}$ children or $4\frac{1}{2}$ children are impossible values for the variable. On the other hand, many variables are considered to have the possibility of occurring at all different values within a specified range and are

† See Stevens [1951, pp. 1–49] for a more complete presentation of this viewpoint on measurement.

‡ Note that whereas the numbers 1 and 2 were chosen to stand for males and females, respectively, this was a perfectly arbitrary choice and any two other numbers could have been chosen. The only restriction is that the two numbers must be different.

called *continuous* variables. Thus, the variable "height of adult males" is considered to be continuous.

Stevens [1951] has presented a system for classifying measurement rules that is useful in applied statistics. He has identified four general types of measurement rules. Each type is known as a *scale of measurement*. If the researcher knows under which scale his measurement rule falls, his choice of statistical methods is limited. This point is very important for the student, since the basic organization of this book is in terms of measurement procedures.

Nominal scales of measurement consist of categories to which numbers may be assigned as codes. For example, the numbers assigned to college courses are convenient labels used in place of the more cumbersome course titles (names) and represent a nominal scale of measurement.† Nominal scales are a very low level of measurement; and in research applications, it is generally the number of occurrences of each category that is recorded. Thus, the tabulation below summarizes data for a dichotomous variable (sex) measured on a nominal scale for a classroom containing a total of 27 children.

Category of the variable	Value	Number of occurrences
Male	1	12
Female	2	15

Ordinal scales of measurement necessarily involve some relationship of "greater-than-ness" or "less-than-ness." That is, values of the variable exist in ranked order. The finishing order of horses in a horse race represents an ordinal scale of measurement. In general, whenever objects, organisms, or events are assigned ranks, the measurement is ordinal in nature.

Interval scales of measurement occur whenever values of the numbers assigned to objects, organisms, or events are chosen so that each unit corresponds to an equal amount of the characteristic being measured. Consider the ordinary Fahrenheit temperature scale for which each 1°F represents a constant amount of heat energy. Thus, a difference of 1°F anywhere along the temperature scale (say, between 42°F and 43°F or between 81°F and 82°F) always stands for exactly the same amount of heat energy. This feature should be

---

† Some colleges and universities use numbering systems that are partially ordinal in nature, since higher-numbered courses (e.g., 100-level courses) are for juniors and seniors only.

compared with the properties of an ordinal scale of measurement. If horse *A* finished first in a race, horse *B* finished second, and horse *C* finished third, you could *not* then assume that horse *A* arrived at the finish line as much ahead of horse *B* as horse *B* finished ahead of horse *C*. In fact, horse *A* may have bettered horse *B* by several seconds and horses *B* and *C* finished only a fraction of a second apart.

Ratio scales of measurement are most typical of measurement involving physical quantities. A ratio scale has not only equal units but also a zero point that is absolute in the sense that 0 represents complete lack of the characteristic being measured. Height of persons in inches or weight in pounds are everyday examples of ratio scales. Note, however, that the Fahrenheit temperature scale is not ratio, for the zero point (that is, 0°F) is not set at the point where there is absolutely no heat energy. On the other hand, the Kelvin, or absolute, temperature scale is a ratio scale, since 0°K is absolute zero.

Stevens' four scales of measurement progress from simple to complex. As one moves from nominal scales toward ratio scales, more and more restrictions are placed on the measurement rule. For example, nominal scales require only unordered categories, but ordinal scales require categories ordered in some definite way. The table below summarizes the essential features of each of the four scales of measurement.

Scale of measurement	Essential features	Example
Nominal	Unordered categories	Political party affiliation
Ordinal	Ordered categories	Ribbons in a talent contest
Interval	Equal units along the scale	Fahrenheit temperature
Ratio	Equal units and absolute zero point	Distance in feet

In this textbook, we use technical terms to refer to the different kinds of data generated when the different types of measurement rules are applied. Thus, *categorical data* is the term used to describe data that result from the application of a nominal scale of measurement; *rank-order data* refers to data derived from the use of an ordinal scale of measurement; and *measured data* refers to data resulting from the use of either an interval or a ratio scale of measurement. This terminology is basic to the organization of the textbook and is summarized in the table below.

Kind of data	Type(s) of measurement rule
Categorical	Nominal
Rank-order	Ordinal
Measured	Interval or ratio

1.3   Other characteristics of research situations

A research study may involve one, two, or more separate groups of subjects. A group of subjects may be either a random sample or a population, depending upon whether the research problem calls for inferential or descriptive statistical procedures. In experimental research, the groups are chosen so that they possess contrasting characteristics with respect to some comparison that is of interest to the researcher. As an example, consider the problem of determining the effects of "anxiety" level upon learning to perform a psychomotor task. The researchers may select two groups (either populations or samples) in such a way that one group represents people of relatively low manifest anxiety and the other people of relatively high manifest anxiety (or, alternatively, the researcher may induce either low or high anxiety in his subjects by suitable pre-experimental instructions, training, etc.). If both groups receive exactly the same amount of training on the psychomotor task under identical training conditions, this would allow for a comparison between the achievement of high- and low-anxiety groups (assuming, of course, that the two groups did not differ systematically with respect to some additional trait such as general intelligence or mechanical ability).

In either a one-group or a two-or-more-group research situation, there may be one or more variables that are analyzed. In the case of the experimental study just described, there is one variable that is of interest in analyzing the results of the study.†   That variable will be the achievement of each subject on the psychomotor task. When the statistical analysis of a research study involves just one variable, the analysis is referred to as *univariate*. The analysis of some

---

† However, there may have been more than one variable involved in the study, since it would be typical to set up the two groups by first administering some form of manifest anxiety test and then selecting individuals with high and low scores to form the two groups.

research studies involves two variables and is called *bivariate* analysis. A common example of a bivariate analysis is one in which two measures of achievement, for example English and mathematics, are obtained for each individual in one group. The analysis may then be used to determine what relationships exist between achievement in the two subject-matter areas in the one group of subjects. Three or more variables may be considered in one statistical analysis by the procedures of *multivariate* analysis. The techniques of multivariate analysis represent relatively advanced procedures and are not treated in this textbook.

1.4 Factors related to the choice of statistical procedures

The topics already discussed in this chapter provide a basis for deciding which statistical procedures are most appropriate for the analysis of data of various sorts. In order to select proper statistical procedures, the research worker must first decide whether his research design involves measurement of an entire population or of a random sample from a population. When a population is involved, the techniques of descriptive statistics are appropriate; when random samples are used, those of inferential statistics are appropriate. Once this decision has been made, the researcher must decide how many (1) variables and (2) groups of subjects are involved in the analysis. This will indicate (1) whether univariate, bivariate, or multivariate analysis is called for and (2) whether statistical procedures for one, two, or more groups are required. Finally, the researcher must examine his variables and classify each of them as either categorical, rank-order, or measured. These four steps will narrow down the possible applicable statistical procedures to a point where relatively few procedures are still appropriate. As an example, consider a study in which data are collected concerning the relationship of the method of study used by the learner to the number of nonsense syllables memorized. More specifically, two random samples of human subjects are drawn from a population of interest. One group memorizes a list of 20 nonsense syllables by method 1, which involves studying the entire list for 20 minutes; the second group studies by method 2, which entails studying each word for 1 minute, until all 20 words are studied. The response variable is the number of nonsense syllables each subject can recall from memory 1 hour after the learning trials.

As a first step in deciding which general class of statistical techniques is appropriate to this study, it should be recognized that samples, not entire populations, are utilized in the study. Hence, inferential statistical procedures are appropriate. To narrow the choice down further, note that the study involves two groups of subjects (method 1 group and method 2 group) and one variable. The variable is the "number of syllables recalled" and this is clearly based on a ratio scale of measurement. Thus, the appropriate statistical techniques for analyzing the results of this study would be univariate analysis by inferential methods for two or more groups with measured data.

This textbook has been organized in terms of decision processes similar to that just described. Chapter 2 presents statistical techniques designed for use in one-group studies with one variable that yields categorical data. Chapter 3 treats techniques applicable when two categorical variables are measured for one group of subjects, or when one categorical variable is analyzed for two or more groups. Within the chapters, subsections are devoted to descriptive and inferential techniques. The organization of the remainder of the book can be seen by studying Table 1.1. By locating the appropriate combination of type of data, number of groups, and type of analysis, the researcher will find listed separately the relevant techniques of descriptive and inferential statistics. The example presented in the preceding paragraph would lead to the inferential techniques listed in cell 18 (i.e., for measured data with two groups of subjects and one variable). The specific names of the techniques (e.g., analysis of variance) are of no special interest at this time; however, the overall organizational scheme should be studied until its rationale is apparent to the user. It should be noted that not all possible research situations can be classified to fit into Table 1.1. (Specifically, no techniques of multivariate analysis have been included; it is assumed that whenever bivariate analysis is utilized, both variables are of the same type—e.g., two categorical variables or two measured variables but not one categorical variable and one measured variable.) Also, the only procedures listed for two or more groups are those for univariate analysis because, as is true for multivariate analysis, those techniques which have been omitted represent advanced topics in applied statistics.

1.5   Conventions with numbers and symbols

The student who feels that he does not have a particularly strong background in mathematics should not be discouraged from studying elementary statistical methods. In fact, understanding the *basic* concepts of elementary statistics requires very little mathematics beyond simple arithmetic. On the other hand, in order to apply statistical techniques for the solution of particular problems, the student will find it necessary to learn certain rules for dealing with numbers and algebraic quantities. In addition, certain special notations are used very often in this book and by research workers, since they provide a convenient

TABLE 1.1   Classification of statistical techniques

Number of groups	Type of analysis	Purpose	Categorical	Rank-order	Measured
ONE	UNIVARIATE	DESCRIPTION	(1) One-way tables; bar, pie diagrams; modal class; conversion to proportions	(7) Array; ranking; cumulative frequencies and proportions	(13) Frequency distribution; histogram; frequency polygon; ogive; percentiles; averages; measures of variability, skewness, and kurtosis
		INFERENCE	(2) Binomial test; chi-square goodness-of-fit test	(8) Kolmogorov-Smirnov goodness-of-fit test	(14) Normal curve theory; z and t tests; chi-square variance test
	BIVARIATE	DESCRIPTION	(3) Two-way tables; comparative diagrams; conditional and marginal proportions	(9) Bivariate arrays; rank-order correlation coefficient	(15) Scatter diagrams; prediction equations; Pearson correlation coefficient; standard error of estimate
		INFERENCE	(4) Chi-square tests of independence and experimental homogeneity	(10) Testing significance of rank-order correlation coefficient	(16) Bivariate normal distribution; testing significance of Pearson coefficient
TWO OR MORE	UNIVARIATE	DESCRIPTION	(5) See cell 3	(11) Comparative tables and cumulative proportions	(17) See cell 13
		INFERENCE	(6) See cell 4	(12) Kolmogorov-Smirnov two-sample test; Mann-Whitney U test	(18) F test of homogeneity of variance; analysis of variance

shorthand. The more commonly used notations and conventions are explained in the sections below. For additional information and for practice in doing arithmetic and algebraic exercises, the student should consult Walker [1951] or Baggaley [1969].

1.6 Algebraic notation

English and Greek letters are often used in statistics to stand for variables, constants, and computed quantities. For example, in a research study, values of the variable Stanford-Binet intelligence-scale scores may be obtained from a population of interest. As a shorthand procedure, the variable may be referred to as the X variable and the individual intelligence scores as X scores. Furthermore, by using a subscript notation, specific intelligence scores can be precisely designated. To illustrate this, consider the following list of intelligence scores for a group of five students. If the individuals are numbered from 1 through 5,

Student number	X score (intelligence)
1	$X_1 = 103$
2	$X_2 = 97$
3	$X_3 = 141$
4	$X_4 = 121$
5	$X_5 = 109$

then a subscript may be used with an X to indicate which student's score is being named. Thus X_3 stands for the intelligence score of student 3 and is equal to 141, and X_5 stands for the intelligence score of student 5 and is equal to 109. In some cases it is useful to be able to name values of a variable before the data are collected or when the specific values of the scores are not known. For example, a variable may be called the Y variable, and the *total number* of scores may be called n. Then the Y scores would be $Y_1, Y_2, \ldots, Y_n$, where the dots indicate that the subscript would increase until it equaled n. For the table of intelligence-test scores given above, n is equal to 5 and the X scores would be written $X_1, X_2, \ldots, X_5$. It is also useful to be able to refer to any of the scores without specifying any particular score. Thus, X_i stands for the ith X

score; but *i* must be given a specific value before it is known just which *X* score is named.

As a final illustration of the use of algebraic notation, consider the expression $dX_1 + dX_2 + dX_3$. If $d = 2$ and if the *X* values are the intelligence-test scores from the table above, then after substitution the expression becomes $2(103) + 2(97) + 2(141) = 206 + 194 + 282 = 682$.

One caution to the student is appropriate at this time. Many statistical procedures and techniques are also referred to by names that contain letters from the alphabet. For example, the *t* test is a technique from inferential statistics.

1.7 *Rounding conventions*†

During computations, such as finding products or quotients, the answers result in more decimal places than the researcher may desire. Also, observations may be recorded to, say, hundredths or thousandths, whereas the researcher may want to have all results expressed as tenths or whole numbers. In such cases, it becomes necessary to *round* numbers. Consider the result when 4.31 is multiplied by 2.16. The product is 9.3096. Since the original numbers each contain only two decimal places, the researcher may desire to report the product to only two decimal places. For this example, the answer would obviously be 9.31 because 9.3096 is closer to 9.31 than to 9.30. As a general rule, a number is always rounded to a number with the desired number of digits that is closer to the original number. A problem arises, however, when the number to be rounded ends in 5, 50, 500, etc. In this case, the number is exactly halfway between two numbers. For example, 21.25 is exactly halfway between 21.2 and 21.3. You may have previously learned to round up when numbers end in 5. This rule is *not* generally applied in statistics because it results in a constant upward rounding error. The rule used states that a number ending 5, 50, 500, etc., should always be rounded so that the rounded number is even. Thus, 21.25 would be rounded to 21.2, if one-decimal accuracy were desired, since 21.2 is even and 21.3 is not.

Whenever a number is rounded, rounding should be completed in one step rather than successive steps. If one-decimal accuracy were desired in the result, then 23.1241 would be rounded immediately to 23.1. If rounding is not done in one step, an error can occur when the number ends in 5, 50, 500, etc., during some stage of rounding. For example, if 1.251 were rounded to one

† This section presents only minimum essentials related to the topic of rounding. Technical Appendix A presents additional concepts that may be studied.

decimal place by successive steps, the results would be 1.251 to 1.25 to 1.2. But this is clearly an error, since 1.251 is closer to 1.3 than to 1.2. The examples below should be studied by the student until the rounding conventions are clearly understood.

Original number	Round to	Result
21.35	One decimal place	21.4
21.25	One decimal place	21.2
14.5001	Whole number	15
18.532	Whole number	19
18.532	One decimal place	18.5
18.5	Whole number	18

In measurement, numbers obtained by counting procedures are referred to as "exact numbers," since only whole, integral values are possible. When measurement involves continuous variables, the resulting numbers are referred to as "approximate numbers," since measurement is never absolutely accurate in these cases. For example, if the length of a table is measured, the result may be reported as 48 inches. However, a more accurate procedure may result in a length of 47.98 inches. If the accuracy of the measuring procedures were further refined, a somewhat different answer might be obtained. The point is that the true length of the table is unknown but by the use of more refined and more accurate measuring procedures, a very good *approximation* to the true length may be obtained. Nevertheless, the measurement will always be in error by some small amount.

1.8   *Mathematical operators*

Everyone is familiar with the mathematical symbols used to indicate, or instruct, certain mathematical operations. Symbols that give instructions or indicate operations to be performed are referred to as *operators*. For example, the $+$ in $8 + 5$ instructs you to add 5 to 8; thus, $+$ represents the "add" operator. Other common mathematical operators are: the subtract operator $-$, as in $8 - 5$; the multiply operator $\cdot$, as in $8 \cdot 5$; and the divide operator $/$, as in $8/5$. In statistics, some additional operators are commonly used. The first of these, the summation operator, is indispensable in statistical work, and its properties are especially important for the student. Suppose that the sum of all numbers in a group is required. For example, on a short quiz, a group of eight students made the following scores: 7, 9, 1, 8, 4, 3, 8, 10. In order to indicate the total sum of these eight scores with the simple addition operator, it would be necessary to write: $7 + 9 + 1 + 8 + 4 + 3 + 8 + 10$. With large groups of numbers, it is inconvenient to write out expressions of this sort. Instead, a more flexible usage can be achieved by first calling the quiz scores the X variable. Then, the first score is X_1, the second score is X_2, and so forth. In general, any one of the scores is X_i, and the total sum of the eight scores can now be written $X_1 + X_2 + X_3 + X_4 + X_5 + X_6 + X_7 + X_8$. Note that the only difference among the terms of this expression is in the value of the subscript. The summation operator capitalizes on this fact in the following way. The uppercase Greek letter sigma, written Σ, is used to indicate summation. Summation of the eight quiz scores can be compactly indicated by

$$\sum_{i=1}^{8} X_i$$

where the term $i = 1$ below the sigma indicates that the first X score to be included in the sum is X_1 and the 8 above the sigma indicates that the last X score to be included in the sum is X_8. In other words, the expression gives the instruction "find the sum of the X scores, starting with X_1 and continuing through X_8." Thus,

$$\sum_{i=1}^{8} X_i = X_1 + X_2 + X_3 + X_4 + X_5 + X_6 + X_7 + X_8$$

To indicate a sum for part of the group of X scores, only minor modifications need to be made. Thus

$$\sum_{i=1}^{4} X_i$$

indicates the sum of just the first four X scores (that is, $X_1 + X_2 + X_3 + X_4$ or $7 + 9 + 1 + 8$) and equals 25. Also,

$$\sum_{i=3}^{5} X_i = X_3 + X_4 + X_5 = 1 + 8 + 4 = 13$$

The summation operator can be applied to more complex expressions than those in the above examples. Consider the expression

$$\sum_{i=1}^{3} (X_i + Y_i)$$

which stands for $(X_1 + Y_1) + (X_2 + Y_2) + (X_3 + Y_3)$; the expression

$$\sum_{i=1}^{3} X_i Y_i$$

which means $X_1 Y_1 + X_2 Y_2 + X_3 Y_3$; and the expression

$$\sum_{i=1}^{4} X_i^2$$

which means $X_1^2 + X_2^2 + X_3^2 + X_4^2$. When powers or exponents appear in an expression with summation operators, it is necessary to be careful in interpreting the expression. For example,

$$\left(\sum_{i=1}^{3} X_i\right)^2$$

means

$$(X_1 + X_2 + X_3)^2 = X_1^2 + X_2^2 + X_3^2 + 2X_1 X_2 + 2X_1 X_3 + 2X_2 X_3$$

This is quite different from

$$\sum_{i=1}^{3} X_i^2$$

which is simply $X_1^2 + X_2^2 + X_3^2$.

As noted earlier in this chapter, the number of scores measured in a study is referred to as n. Thus,

$$\sum_{i=1}^{n} X_i$$

stands for $X_1 + X_2 + \cdots + X_n$, where the summation is continued over all n scores.

Some elementary rules for using the summation operator are presented below.

RULE 1 If only addition or subtraction is indicated between two or more terms, the summation operator may be applied separately to each term.

For example, in the expression

$$\sum_{i=1}^{n} (X_i + Y_i)$$

the X and Y scores are connected by addition. Therefore,

$$\sum_{i=1}^{n} (X_i + Y_i) = \sum_{i=1}^{n} X_i + \sum_{i=1}^{n} Y_i$$

Similarly,

$$\sum_{i=1}^{n} (X_i - Y_i + Z_i) = \sum_{i=1}^{n} X_i - \sum_{i=1}^{n} Y_i + \sum_{i=1}^{n} Z_i$$

However, this rule would not apply to

$$\sum_{i=1}^{n} (X_i + Y_i)^2$$

until squaring of the expression in parentheses had been completed. Thus,

$$\sum_{i=1}^{n} (X_i + Y_i)^2 = \sum_{i=1}^{n} (X_i^2 + 2X_iY_i + Y_i^2)$$

$$= \sum_{i=1}^{n} X_i^2 + \sum_{i=1}^{n} 2X_iY_i + \sum_{i=1}^{n} Y_i^2$$

RULE 2   The summation operator applied to a constant is equivalent to multiplying the constant by the number of terms implied by the upper and lower limits shown on the summation operator.

To clarify this rule, consider

$$\sum_{i=1}^{3} 8$$

This is equivalent to $8 + 8 + 8$ or $8 \cdot 3$, since i goes from 1 through 3, and indicates a sum of three terms. In general, for m less than or equal to n,

$$\sum_{i=1}^{n} b = nb \qquad \text{and} \qquad \sum_{i=m}^{n} b = b(n - m + 1)$$

Thus,

$$\sum_{i=1}^{n} 6 = 6n \qquad \text{and} \qquad \sum_{i=4}^{9} 8 = 8(9 - 4 + 1) = 8 \cdot 6 = 48$$

RULE 3   A constant multiplicative term following a summation operator may be removed and made a constant multiplier preceding the summation operator.

For example,

$$\sum_{i=1}^{n} 2X_i$$

involves the constant multiplier 2 for each X score and can be rewritten

$$2 \sum_{i=1}^{n} X_i$$

Similarly,

$$\sum_{i=1}^{n} 2X_iY_i = 2 \sum_{i=1}^{n} X_iY_i$$

represents the constant multiplier 2 for each XY product.

The following examples illustrate various combinations of rules 1 through 3:

$$\sum_{i=1}^{n} (X_i + d) = \sum_{i=1}^{n} X_i + \sum_{i=1}^{n} d = \sum_{i=1}^{n} X_i + nd$$

$$\sum_{i=1}^{n} (2X_i + 2d - 4) = \sum_{i=1}^{n} 2(X_i + d - 2) = 2 \sum_{i=1}^{n} (X_i + d - 2)$$

$$= 2 \sum_{i=1}^{n} X_i + 2nd - 4n$$

$$\sum_{i=1}^{n} (X_i - Y_i)^2 = \sum_{i=1}^{n} (X_i^2 - 2X_iY_i + Y_i^2)$$

$$= \sum_{i=1}^{n} X_i^2 - 2 \sum_{i=1}^{n} X_iY_i + \sum_{i=1}^{n} Y_i^2$$

Whereas the summation operator is the most important one for the new student of statistics to study and understand, several additional special operators will be encountered from time to time. In some research work, either positive or negative scores may occur. For example, if a group of subjects is administered the same test on two different occasions and if the variable in the experiment is assumed to be the difference between scores on these two occasions, then it is possible that some subjects will show negative differences. For example, a student who scores 51 on the first administration of the test and 58 on the second administration, shows a difference of $+7$ points (or a 7-point gain); however, a student scoring 41 on the first administration and 38 on the second administration, shows a difference of -3 points (or a 3-point loss).

In some statistical applications, it is desirable to ignore negative signs and consider all numbers to be positive. That is, a positive gain of 3 points or a negative gain of 3 points is treated identically. In order to indicate that negative signs should be ignored, the *absolute-value* operator may be used. Thus, $|b|$ is read "the absolute value of b." For example, $|21| = 21$, and $|-21| = 21$. Note that after the absolute-value operator is applied, the resulting number is always positive. Whenever the absolute-value operator is applied to an expression, the operations indicated in the expression should be performed before the absolute value is taken.

Another operator of interest in statistics is the *factorial* sign !. The factorial sign indicates the continued product of all positive integers equal to or less than the number to which the operator is being applied. In other words, $n! = n(n - 1) \cdots 1$. As an example, $5! = 5 \cdot 4 \cdot 3 \cdot 2 \cdot 1 = 120$, and $10! = 10 \cdot 9 \cdot 8 \cdots 1 = 3,628,800$. Also, by definition, $0! = 1$.

The last operator to be discussed here is the multiplication operator. The upper case Greek letter pi, written Π, is used to indicate multiplication of all numbers in a group. For example,

$$\prod_{i=1}^{3} X_i = X_1 X_2 X_3$$

and

$$\prod_{i=1}^{n} (X_i + Y_i) = (X_1 + Y_1)(X_2 + Y_2) \cdots (X_n + Y_n)$$

Caution should be observed in generalizing the rules from the summation operator to the case of the multiplication operator. Thus,

$$\prod_{i=1}^{n} b = b^n \quad \text{and} \quad \prod_{i=1}^{n} bX_i = b^n \prod_{i=1}^{n} X_i$$

Finally

$$\prod_{i=1}^{n} (X_i + Y_i)$$

does *not* equal

$$\prod_{i=1}^{n} X_i + \prod_{i=1}^{n} Y_i$$

and, in fact, no simple rule applies here.

In addition to the special operators discussed above, the student should also learn the mathematical symbols commonly used to express *relationships* between two numbers or two expressions. Everyone is familiar with the equality sign $=$, which states that two numbers or expressions stand for the same quantity. If two numbers or expressions are not equal, this can be symbolized by the general inequality sign $\neq$. Thus, $5 = 3 + 2$ and $8 + 2 = 2 \cdot 5$, but $2 \neq 3$ and $5 + 7 \neq 4 + 4$. When two numbers or expressions are not equal, then one of the numbers or expressions must have a larger value than the other. This relationship can be expressed by the specific inequality sign $>$, which stands for "greater than," or by the sign $<$, which stands for "less than." For example $3 > 2$ is read "3 is greater than 2." This fact can also be symbolized as $2 < 3$, or "2 is less than 3." In using these inequality signs, note that the smaller number or expression is always written on the same side as the point of the inequality sign: larger number $>$ smaller number, or smaller number $<$ larger number. The specific inequality signs may be combined with the equal sign to yield symbols for two other relationships. Thus, $X \geq Y$ states that X is either equal to or greater than Y, and $5 \leq 8$ is read "5 is equal to or less than 8."

With the completion of the material concerning mathematical conventions and special operators, the student should be prepared to attack the discussions of the procedures of applied statistics presented in the remainder of this book. It is recommended that the student review the material in Sec. 1.4 before continuing his study of the text.

Problems

1.1.   Define the following terms:

   a. Population
   b. Descriptive statistics
   c. Sample
   d. Inferential statistics
   e. Measurement rule
   f. Variable
   g. Discrete variable
   h. Continuous variable
   i. Four scales of measurement
   j. Univariate
   k. Bivariate
   l. Multivariate
   m. Absolute value
   n. Mathematical operator
   o. Dichotomy

1.2.   For each variable, indicate whether it is continuous or discrete.
   a. Weight of watermelons
   b. Wattage of light bulbs
   c. Number of pills in a bottle
   d. Speed of a race car
   e. Number of students in a class
   f. Thickness of hand-dipped candles

1.3.   A variable which is *not* discrete is:
   a. Class size
   b. Golf score
   c. Distance to New York City
   d. Make of automobiles

1.4.   A scale of measurement is a system of:
   a. Tests
   b. Rules
   c. Units
   d. Characteristics

1.5. Nominal measurement corresponds to:
   a. Ordering
   b. Assigning
   c. Categorizing
   d. Scaling

1.6. Round each number to one decimal place.
   a. 27.15
   b. 45.65
   c. 6.53
   d. 24.453

1.7. Round each number to a whole number.
   a. 24.5
   b. 10.51
   c. 13.57

1.8. Define the following symbols.

   a. !
   b. $>$
   c. $<$
   d. $=$
   e. $\neq$
   f. $\geq$

   g. $\leq$
   h. $|\ |$
   i. Σ
   j. X_i
   k. Π

1.9. a. Solve for the value of

$$\sum_{i=1}^{8} i$$

   b. Write the following in expanded form.

$$\sum_{i=2}^{4} Y_i$$

   c. Write the following in expanded form.

$$\sum_{i=1}^{3} X_i Y_i$$

   d. Simplify

$$\sum_{i=1}^{4} (dX_i + c)$$

   e. Write in expanded form and solve for the value of 5!

1.10. List four factors that are relevant to the choice of appropriate statistical techniques for the analysis of a set of data.

Two

descriptive and inferential methods for univariate categorical data

2.1 *Descriptive procedures*

Tabular and graphical techniques

The simplest types of practical questions that may be answered by statistical methods are: "How many?" or "How often?" These questions occur commonly in everyday life and many people with no formal training in statistics are able to carry out the appropriate procedures to answer them. Essentially, such problems involve counting, or *enumerating*, the frequency of occurrence for cases falling into each of two or more categories, and then summarizing these results in the form of a table or graph. In general, summaries of data for populations are called population *distributions*. Table 2.1 displays the results of an enumeration based upon the sexes of 10 children in a classroom. The data upon which Table 2.1 is based are clearly derived from a nominal scale of measurement, since the variable, sex, represents a set of unordered categories. Also, descriptive statistics are appropriate to this situation, since the entire population of interest consists of these 10 children. The actual process of measurement for this simple situation can be represented in two ways. (1) Data are collected in *tally* form after classifying a child as being either male or a female (the second column of Table 2.1 shows the result of tallying). (2) A code number is assigned to each individual in the classroom in order to adhere strictly to the definition of measurement on a nominal scale. If the number 1 stands for male and the number 2 for female, then the results of this assignment would

TABLE 2.1   Frequency distribution for sexes of 10 children

Sex	Tally	Frequency
Male	ꟷ꒐ꟷ	6
Female	////	4
Total		10

TABLE 2.2   Distribution of sex for 10 children

Student number	Category	Code	Student number	Category	Code
1	Male	1	6	Female	2
2	Male	1	7	Female	2
3	Female	2	8	Male	1
4	Male	1	9	Male	1
5	Female	2	10	Male	1

appear as in Table 2.2. Table 2.1 can easily be derived by counting the number of 1s and of 2s that occur in Table 2.2. This second procedure involves more work than the first; and in actuality, the tallying procedure is more commonly used. However, the second procedure does more clearly expose the underlying process of measurement that results in categorical data.

Tables such as Table 2.1 are referred to as *one-way classification tables* and may be translated into several different types of graphs. Figure 2.1 shows the sex data for the 10 students represented as a *bar graph*, or *bar diagram*. The distinctive features of a bar graph are that (1) the frequency for each category is represented by the height of a rectangle, or bar; and (2) the values, or categories, of the variable are represented at arbitrary, but usually equal, distances

4

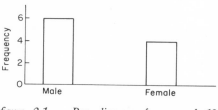

figure 2.1   Bar diagram for sex of 10 children.

along a horizontal axis.   To illustrate these techniques for the case of a variable with more than two values, let us consider the enumerative data showing fields of specialization for a population of 75 graduate psychology students enrolled in one department of psychology (Table 2.3).   This table can be used to construct

TABLE 2.3   *Frequency distribution for 75 graduate psychology students*

Major	Frequency
Experimental	21
Clinical	18
Social	8
Quantitative	6
Industrial	7
Educational	15
Total	75

a bar graph (Fig. 2.2), or can be translated into some other equivalent graphical form.   A commonly used alternative to the bar graph is the *pie diagram*, in

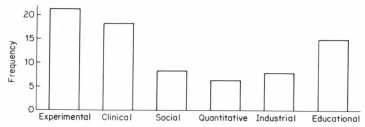

figure 2.2   Major areas for 75 graduate psychology students.

which each frequency is converted into a proportionate sector of a circle (Fig. 2.3). The circumference of a circle subtends a total angle of 360° and the pie diagram is constructed by dividing these 360° into sectors that are proportional to the frequencies for the categories. For the data in Table 2.3, the total frequency is 75 and the total angle of the circle, 360°, is thus taken to represent these 75 cases. Then, starting at any arbitrary point on the circle, an arc is measured off that is proportional to the frequency for the experimental psychology majors. The arcs for the remaining categories may be found by the same process. For example, the 18 clinical psychology majors represent 24 percent of the total group (i.e., $100(18/75) = 24$) and 24 percent of 360° is 86.4°. Figure 2.4 shows the pie diagram with just the first two sectors marked off.

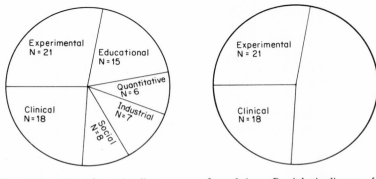

figure 2.3   *Complete pie diagram for psychology majors.*   figure 2.4   *Partial pie diagram for psychology majors.*

Actual marking off of the appropriate arcs on the circle involves the use of a protractor.

It should be noted that a one-way classification table, a bar graph, and a pie diagram each present the reader with exactly the same information: they represent *alternative* methods for displaying distributions of categorical data. That is, the researcher would choose *one* of these procedures, and this choice depends upon both the researcher's personal preference and the nature of the audience for whom the data are intended. Thus, a pie diagram or bar graph may be more appropriate for an audience of persons completely untrained in statistics, since the form of the figure allows for easy comparisons among the numbers in each category. On the other hand, the one-way classification table is easier to construct and more useful to statistically trained persons.

Derived measures

Tabular and graphic procedures preserve data in their original, or nearly original, form. When the number of cases is large, it may be more convenient to sacrifice complete reporting of the data and use *derived measures*. That is, rather than present the data in detail, the researcher may choose to perform certain arithmetic processes on his data to either (1) simplify the form of the data and thereby make it easier to understand or (2) reduce the total amount of data reported. An example involving changing the form of data is that of reporting percentages or proportions† in place of frequencies. Table 2.4 pre-

TABLE 2.4 *Percentage and proportion distribution of psychology majors*

Major	Percentage	Proportion
Experimental	28.0	.280
Clinical	24.0	.240
Social	10.7	.107
Quantitative	8.0	.080
Industrial	9.3	.093
Educational	20.0	.200

sents the psychology data as both percentages and proportions. The use of percentages or proportions enables the user of the data to make certain interpretations more readily. For example, inspection of the percentages from Table 2.4 shows that experimental, clinical, and educational psychology majors constitute nearly three-quarters of the total graduate student group and that the remainder of the students are more or less equally divided among social, quantitative, and industrial psychology. The ease of interpretation of data presented in the form of percentages or proportions provides an especially attractive alternative when communicating results to statistically untrained persons.

The reduction of the total aggregate of data to one or more derived measures is an important concept in statistics. Most users of this book are already familiar with this idea, since they have computed the average‡ of a set of scores by summing the scores and then dividing this sum by the number of scores. Thus, a teacher may record scores from five or six tests of equal length for a

---

† Percentages and proportions are equivalent but utilize different conventions. A percentage involves a comparison based upon 100, whereas a proportion is based upon a comparison with 1.00. Thus 38 percent is equivalent to the proportion .38.

‡ The term "average" is used in a generic sense in statistics, and the type of average described above is known as the *arithmetic mean*. The arithmetic mean will be discussed in Chap. 4.

student during a semester and then find the average of these as the student's final score. Note that the teacher is reducing the total amount of data to just one number, or derived measure, that represents the data.

The average commonly used with categorical data is the *modal class*. The modal class is the category of the variable that has the largest frequency of occurrence. In the case of the psychology-major data, the modal class is experimental psychology, since this category contains 21 graduate students and no other category shows a frequency this large. Occasionally, there may be two or more categories that are tied for having the largest frequency, in which case, no single modal class can be designated. Data are described as being *bimodal* if two categories are tied, *trimodal* if three categories are tied, and *multimodal* if more than three categories are tied for most frequently occurring. For data with just one modal class, this class represents the category in which an observation is most likely to be found, and it is in this sense that the modal class represents an average value. Thus, if you were told *only* that a certain individual was one of the 75 graduate psychology students summarized in Table 2.3, then your best guess would be that he specialized in experimental psychology, since this is the most frequently occurring major.

The use of descriptive measures is not restricted to cases in which only one population is involved. A *comparative* one-way classification table or a *comparative* bar diagram may be constructed to represent the same variable for two or more different populations. Similarly, frequencies for these populations could be converted to percentages or proportions and the modal class for each population identified. Table 2.5 displays data for psychology majors in three depart-

TABLE 2.5   Distribution of psychology majors in three departments of psychology

Major	Frequencies			Percentages		
	1	2	3	1	2	3
Experimental	21	9	29	28.0	22.5	18.1
Clinical	18	11	61	24.0	27.5	38.1
Social	8	10	15	10.7	25.0	9.4
Quantitative	6	4	11	8.0	10.0	6.9
Industrial	7	6	6	9.3	15.0	3.8
Educational	15	0	38	20.0	0.0	23.8
Total	75	40	160	100.0	100.0	100.1

ments of psychology in three different universities. Since the total number of graduate students varies from population to population, comparisons among the groups are simplified if percentages are used in place of frequencies. The percentage data are shown along with the frequencies in Table 2.5. To construct a comparative bar diagram, the percentages rather than frequencies would be used (Fig. 2.5).

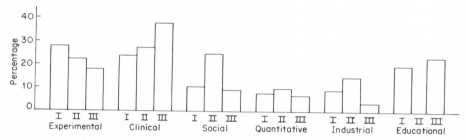

figure 2.5   Comparative bar diagram for students in three departments of psychology.

In summary, univariate categorical data may be represented in tabular form either by showing frequencies in a one-way classification table or by deriving percentages (or proportions) from these frequencies and setting up a table of percentages (or proportions). On the other hand, the same data may be displayed as either a bar diagram or a pie diagram. If only an average value is sufficient for the researcher's purposes, then the modal class may be used. When more than one population is involved, data for these may be summarized in the form of a comparative one-way classification table or a comparative bar diagram.

2.2 *Inferential procedures*

Sampling techniques

It was pointed out in Chap. 1 that whenever observations on an entire population of interest are not obtained by the researcher, it may still be possible to make *inferences* concerning characteristics of the population if random samples from the population are available. In the paragraphs below, discussion of the procedures of random sampling and some extensions of these procedures are presented.

Up to now we have used the term "population" to refer to the total group of objects, organisms, or events that are of interest to the researcher. Strictly speaking, however, the population consists of the *measurements* themselves, not the objects being studied. For example, a population may be defined as the *intelligence scores* of all ninth-grade public-school students in Madison, Wisconsin. Nevertheless, it is common for statisticians to refer to *either* the measurements (e.g., intelligence scores) *or* the objects being studied (e.g., students) as the population. Thus, in this book, population will be used in both these senses.

In order for a sample to be described as random, it is necessary that the sample be drawn from the population in such a manner that each member of the population has an *equal* and *independent* chance of being included in the sample. Note that it is the *process of selecting* the sample that determines whether or not it is random. The requirement that the population members each have an equal chance of being included in the sample is interpreted to mean that no individual or group of individuals is placed in either a more- or less-favorable position than the others. Thus, if one person were drawn randomly from a population of 10 people, this implies that each person has 1 chance in 10 of being the person selected. The requirement that the population members have an independent chance of being included in the sample means that the selection of one person does not influence the selection of any other person in the population. It is easiest to clarify these points by considering the examples below for which the requirements of random sampling are violated.

1)   An instructor in an educational psychology class uses the students in his class as subjects in an experiment and then wishes to generalize his results to all students of the same type in the United States. His sample is clearly not selected in a random fashion, since students not in his class have no chance to be in the sample. Also, the students obviously are not selected independently of one another.

2)   A sociologist is making a survey in a city of 10,000 people. He wishes to include 100 people in his survey so he selects each hundredth name in the local telephone directory and includes that person in his sample. This is not a random sample from the population of the city, since it excludes everyone who is not listed in the telephone directory and it increases the chance of inclusion for anyone with two or more listings in the directory. Also, the choices are not independent, since once the first name is chosen, the rest are automatically determined (i.e., each hundredth name following the first is included and all others are excluded).

It is not unusual for researchers in the behavioral sciences to select samples in a manner similar to the examples above and then to apply the techniques of inferential statistics to these measurements. This appears to be by far the most common weakness in the use of statistical procedures in the behavioral sciences. Whenever it is not practical to use random sampling from the entire population of interest, the researcher is on safer grounds if he redefines his population in some narrower sense, randomly samples from this redefined group, and then restricts his conclusion to this group. For example, rather than attempt to generalize to all college students in the United States, a researcher might redefine his population of interest as all students at one particular college. While it may be impractical to select randomly from the nationwide group, local random sampling at one college may be a feasible plan.

The researcher who has available a list of names of all persons in his population of interest and who then selects each tenth, each hundredth, or some other regular pattern of individuals, is displaying an inexcusably lazy attitude toward his sampling. To ensure random sampling, he should avail himself of a table of random numbers and base his sampling upon this table. Table F.9 † is a table of random numbers. This table has been specially constructed so that the digits are in nearly random order. To illustrate the use of the random-number table, let us consider the problem of selecting 25 names from a list of 300 names. (The list might be a roster of teachers in a school district and the problem to sample 25 of these who will be interviewed concerning their opinions on local educational issues.) The names on the list are considered to be num-bered from 000 through 299. Thus, we want to select 25 three-digit numbers from the random-number table and then the sample will consist of the 25 individuals so numbered on the list. To ensure complete randomness, the table itself should be entered randomly. Note that the rows of the random-number table are numbered from 1 through 200 and each row contains 50 digits grouped in sets of 5 digits. Thus, the table contains a total of 10,000 digits. A con-venient way of entering the table is to use the serial number on a dollar bill to provide the appropriate row and column of the table. The first three numbers of the serial number determine the row (you must, of course, use a dollar bill whose serial number begins with a three-digit number of 200 or less) and the next two digits determine the column. Therefore, if your dollar bill happens to have the serial number 01426640, you would enter the random-number table at row 14 and column 26. The entry in Table F.9 at the intersection of row 14 and col-umn 26 would be taken as the first digit of the first three-digit random number. This number is 055; the remaining 24 random numbers can be found by reading down the column. If the same number repeats, it is ignored because we do not wish the same person to appear twice in the sample. (This is known as "sam-pling without replacement," since once a person has been sampled, he is removed from the population and is no longer available for sampling.) Also, if a number larger than 299 is encountered, it is skipped, since our population has been numbered from 000 through 299. When the bottom of the table is

† All tables with an "F" number will be found in Appendix F at the end of this book.

reached and additional numbers are still needed, one simply returns to the top of the table and starts at column 29. The student should confirm that if one starts sampling at row 14 and column 26, the required random numbers are as shown in the following table.

055	003	210	198	234
038	243	213	088	230
277	236	290	025	112
268	278	096	240	004
032	151	178	249	254

Certain variations of the random-sampling plan can be used in special circumstances. If the population of interest is divided into subgroups, or *strata*, the technique of stratified random sampling may be used to increase the precision

	Native-born	Foreign-born
Upper class	Native-born Upper class	Foreign-born Upper class
Middle class	Native-born Middle class	Foreign-born Middle class
Lower class	Native-born Lower class	Foreign-born lower class

figure 2.6   Substrata for a population of school-age children.

of a research study. The strata may be considered as subpopulations. This plan is most valuable when the researcher is interested in obtaining results for each stratum separately. For example, a social psychologist may be studying attitudes toward money in a population of school-age children. It is reasonable to assume that the various socioeconomic classes, or strata, would hold differing attitudes toward money. Also, the researcher may be interested in seeing whether or not attitudes were different for children of native-born and children of foreign-born parents. Hence, he might divide the population into both socioeconomic and birthplace strata prior to sampling. The population can be represented as in Fig. 2.6. For each substratum identified in this way, the researcher would draw a separate random sample from among the individuals falling into that substratum. Usually the number of individuals in the sample from each stratum or substratum is chosen to be proportional to that stratum's total size. Thus, if 30 percent of the entire population in the above example were native-born and middle-class, then 30 percent of the total sample would be randomly selected from that substratum. Sampling from the remaining substrata would follow the same plan. In the total sample selected in this manner, each substratum receives the same weight as in the population.

Binomial distribution

The most elementary type of situation to which methods of statistical inference can be applied is that involving a single dichotomized variable. The population of interest for this case consists of objects, events, or organisms that come in two different forms. When a random sample is drawn from a population of this type each member of the sample will represent one of the two possible values of the variable. The term *binomial distribution* is used to refer to a distribution of possible different outcomes that can occur in sampling from a dichotomized population. If we symbolize the two values of the dichotomized variable as A and B, then the probability of A, π_A, is defined as the proportion of times A appears *in the population*. Similarly, the probability of B, π_B, is the proportion of times B appears *in the population*. Of course, for situations in which only a random sample from the population is available, the values of π_A and π_B cannot be computed. If a random sample of size n is chosen, and n_A of the individuals in the sample have the characteristic A, then the sample proportion is $p_A = n_A/n$ for A and, similarly, $p_B = n_B/n$ for B. As will be seen, p_A and p_B are appropriate estimates of the corresponding population values π_A and π_B. Although π_A and π_B are unknown, it is often possible to make reasonable and pertinent *assumptions* concerning the values of π_A and π_B. Inferential techniques can then be used to determine how tenable these assumptions are in light of the data from a random sample.

The last two sentences are extremely important for an understanding of the general approach used in inferential statistics. In general, pertinent assumptions, or *hypotheses*, are made about the nature of a population. These assumptions, or hypotheses, are made in such a way that it is then possible to use the sample data in order to perform statistical *tests* from which the tenability or plausibility of these hypotheses can be evaluated. The test is performed on the basis of data derived from a random sample of the population.

Consider the example of a population defined as all babies born during 1 month in a city. This may be viewed as a binomial population with respect to the sex variable, since each baby can be classified as either male or female. The probability of a male in this population is defined as $\pi_A = N_m/N$, where N_m is the number of males in the population and N is the total number of babies.

If N is 400, and 216 of these are males, then $\pi_A = .54$; similarly, $\pi_B = .46$ is the probability of a female. A probability defined in this way can be interpreted as the likelihood that a single randomly drawn individual from the population will have the specified characteristic. Thus, the likelihood that a randomly drawn baby from the above population will be a male is .54. That is, 54 out of each 100 babies drawn are expected to be males. However, a random sample may, in fact, deviate from this expectation. Thus, it is possible that 55, 56, or more of the babies may be males, or that 53, 52, or fewer are males. Although it is very unlikely, a random sample of 100 babies might consist entirely of males. The fact that random samples do not always contain exactly the expected number of each value of a variable is referred to as *sampling variability*. On the basis of sampling variability alone, it is possible for random samples of 100 babies to be all males or all females, and all intermediate combinations of numbers of males and females are also possible. Of course, extreme outcomes are less likely to occur than are outcomes with about 54 males and 46 females. In the following paragraphs, we will consider methods for actually computing the likelihood, or probability, that a given sample will arise by random sampling from a binomial population.

The example of the newborn babies is not a realistic one for the application of inferential statistics because the proportions of males and females in the population were known. Hence, there is no need to use a random sample from this population to make inferences about these proportions. Consider, however, a research project on ESP (extrasensory perception). A researcher may have each of 20 randomly chosen college students guess the suit of an ordinary playing card that is hidden from their view. Although we do not know what proportion of the population (i.e., all students at the college) could guess the suit of the card correctly, it is possible to make a pertinent assumption concerning this proportion. If we hypothesize that no ESP ability exists in the population and that each student simply makes an independent guess concerning the card's suit, then it is reasonable to assume that one in four, or 25 percent, of the students in the population could guess the suit of the card correctly because there are four suits to choose among. If we symbolize a correct guess as A, then

$$\pi_A = N_A/N = .25$$

is the mathematical statement of our assumption. An assumption of this type is commonly referred to as a *null hypothesis*† and is written H_0. Thus, a compact form of stating the above assumption is $H_0: \pi_A = .25$. Our assumption also implies that $\pi_B = .75$ is the probability of making an incorrect guess.

† The use of the term "null hypothesis" is based on the fact that such a hypothesis can generally be written in a form which implies *no difference* between two quantities. Thus, $H_0: \pi_A = .25$ could be written in the equivalent form $H_0: \pi_A - .25 = 0$.

For a binomial population, the probabilities of the two values of the dichoto-mized variable must obviously sum to 1. For this example, $\pi_A + \pi_B = 1.00$. From the null hypothesis, it is possible to determine the probability that a specific random sample will occur. Under the null hypothesis, we would *expect* 5 of the 20 individuals in the sample to guess correctly the suit of the card. Owing to sampling variability, the actual number may deviate from five; the problem with respect to a single sample is thus to decide either (1) that the deviation is due to sampling variability alone or (2) that the null hypothesis is untenable and that the individuals are not simply guessing at random. That is, we might decide that some real ESP ability existed in the population. Mathematically, alternative 2 can be represented by the inequality $\pi_A \neq .25$. More specifically, in this example we would be interested in the alternative $\pi_A > .25$, since only this may be interpreted as implying ESP ability in the population.

In general, the decision that a null hypothesis is tenable and that deviations from expectation are due to sampling variability is referred to as "accepting the null hypothesis." Deciding that the null hypothesis is untenable is called "rejecting the null hypothesis." Of course, accepting or rejecting the null hypothesis does not in any way affect the reality of H_0 being either true or not true. It is important to realize that a researcher may be in error when he accepts *or* when he rejects a null hypothesis. In the ESP example, we might *decide* on the basis of a test of inference that H_0 is untenable. However, H_0 may still be true *in fact*. The error of rejecting a null hypothesis which is actually true is called a *Type I*, or *alpha* (α), *error*. When a researcher accepts a null hypothesis which is in fact not true, he is committing a *Type II*, or *beta* (β) *error*. Of course, in any specific case, the researcher cannot tell whether or not his decision represents an error. However, it will be seen that he can directly control the *probability* of making a Type I error and he can influence the *probability* of a Type II error by indirect means. Table 2.6 summarizes these concepts concerning decision making and errors; we also indicate the complementary outcomes which represent correct decision. For example, if the null hypothesis is, in fact,

TABLE 2.6   *Decision concerning null hypothesis*

		Accept	Reject
In reality, null hypothesis is:	True	Correct decision	Type I (α) error
	Not true	Type II (β) error	Correct decision (power)

true and our decision is to accept H_0, this is a correct decision.   Also, if the null hypothesis is not true and our decision is to reject the hypothesis, this, again, is a correct decision.   The likelihood that a statistical test will correctly reject false null hypotheses is referred to as the *power* of the test and this concept is discussed in more detail later in this chapter.

Some statisticians prefer to avoid the phrase "accept the null hypothesis" and say instead "fail to reject the null hypothesis."   That is, the data are not sufficiently unusual to cast doubt upon the validity of the null hypothesis; however, ultimate judgment is reserved, since future, more sensitive research may reveal the untenability of the null hypothesis.   In this book, we shall talk about "accepting" a null hypothesis because statistical hypothesis testing procedures do pose a choice between the null hypothesis and the alternate hypothesis. Nevertheless, we recognize that the acceptance of a null hypothesis may be upset by future evidence.   Thus, "accepting a null hypothesis" is taken to mean that the available data do not contradict the null hypothesis.

Let us return to the ESP research and see how the null hypothesis can be tested on the basis of results from the random sample of 20 students.   Suppose that 8 of the 20 students were successful in guessing the suit of the hidden card. The problem is to decide whether this result represents a sampling variability from a population in which $\pi_A = .25$ or whether it is more reasonable to assume that $\pi_A > .25$ and that some ESP ability is present in the population.   More specifically, it is necessary to compute the probability of drawing a random sample with the proportion of correct guesses, $p_A = .40$ or greater, from a population with $\pi_A = .25.$† The phrase "$p_A = .40$ or greater" is used rather than "$p_A = .40$" because it is generally of interest to decide how rare or unusual the actual outcome is.‡   Thus, we will compute the probability of obtaining the observed *plus all more extreme* outcomes, that is, the probability of 8 correct guesses *plus* the probability of 9 correct guesses *plus* the probability of 10 correct guesses and so forth through 20 correct guesses.   If we let $P(X)$ represent the probability of X correct guesses and $P(X \geq k)$ represent the probability of k *or more* correct guesses, then the above probability can be written

$$P(X \geq 8) = P(8) + P(9) + P(10) + \cdots + P(20)$$

As will become apparent, the actual computation of $P(X \geq k)$ is a laborious task by hand computation.   However, a procedure involving a theoretical distribution known as "chi square" is available for finding the *approximate* value

---

† We let p_A represent the proportion of correct responses in the sample, whereas π_A represents the proportion expected if the students guess.

‡ This usage is consistent with that necessary when the measured variable is continuous. For continuous variables, the probability of any given sample is usually 0.   However, the probability of the given *or* more extreme samples can be computed.

of $P(X \geq k)$ and similar probabilities.   Chi-square distributions are considered in the next subsection.   The general procedure for computation with a binomial distribution is presented here in order to clarify the reasoning and methods involved.   Also, the exact procedure is useful in cases with small samples.

Consider, first, finding the probability of the most extreme possible outcome in the ESP experiment.   If all 20 students in the sample guess the suit of the card correctly, it is necessary that the first student guess correctly *and* the second student guess correctly *and* the third student guess correctly *and* so forth through the twentieth student.   Under the null hypothesis, each student has a probability of .25 of guessing correctly.   Also, under H_0 each student's guess represents a single *binomial event* with $\pi_A = .25$ and $\pi_B = .75$.   Then, the outcome that all students guess correctly involves the occurrence of successive binomial events each representing a correct guess.   Whenever a sample involves successive binomial events, the probability of this sample can be found by successive applications of the following rule:

PROBABILITY RULE 1   If A and B represent alternative binomial outcomes for which the probability of A is π_A and the probability of B is π_B ($\pi_A + \pi_B = 1$), then the probability of any two successive outcomes is the *product* of the probabilities of the individual events.   Thus, $P(A,B) = \pi_A \pi_B$, $P(A,A) = \pi_A^2$, $P(B,B) = \pi_B^2$, and $P(B,A) = \pi_B \pi_A$.

When rule 1 is applied to determine the probability of all 20 students guessing the suit of the card, this means that we must multiply the probabilities for the individual binomial events.   Since there are 20 such events and each has a probability of .25, the result is $P(20) = (.25)^{20}$.   This obviously is an extremely small value.

Now consider computing $P(19)$.   For this case, 19 students would guess correctly while one student guessed incorrectly.   Suppose that the first 19 students guess correctly and the twentieth student guesses incorrectly.   This outcome could be represented as:

Student	1	2	3	4	5	6	$\cdots$	19	20
Outcome	A	A	A	A	A	A	$\cdots$	A	B
Probability	.25	.25	.25	.25	.25	.25	$\cdots$	.25	.75

The probability is $(.25)^{19}(.75)$.   However, this is not the only possible way that 19 correct guesses could occur.   There are, in fact, 20 different ways that just 19 out of 20 students could guess correctly.   This is so because *any one* of the students could guess incorrectly while the remaining 19 guess correctly. For cases in which an outcome can occur in more than one way, the following rule applies:

PROBABILITY RULE 2   If an outcome can occur in two or more distinct ways, the probability of that event is the *sum* of the probabilities of the distinct outcomes.†

Thus, $P(19)$ is the sum of the probabilities of the 20 different ways in which 19 correct and 1 incorrect guess can be made. Since each of these individual probabilities is $(.25)^{19}(.75)$, the sum of 20 such probabilities is equivalent to $20(.25)^{19}(.75)$; that is, $P(19) = 20(.25)^{19}(.75)$. The multiplier 20 is referred to as a *binomial coefficient*.

For 18 correct guesses, a similar line of reasoning applies. The probability of any specific order of 18 correct and 2 incorrect guesses is $(.25)^{18}(.75)^{2}$. It remains only to determine in how many different ways this outcome can occur. Unfortunately, this is not as simple as in the previous cases. However, a general method is available for computing binomial coefficients for any number of corrects and incorrects. For our example, this involves determining the number of different ways in which 18 events can be selected from a total of 20 events, called the number of *combinations* of 20 events taken 18 at a time. If you imagine a row of 20 holes, then the number of combinations above represents the number of different ways 18 identical pegs could be placed in these 20 holes.‡ Notice that this is equivalent to the number of different ways in which two holes could be left empty. In general, if there is a total of n objects, the number of different combinations of these n objects taken r at a time is symbolized

$$_nC_r$$

and the formula for computing is

$$_nC_r = \frac{n!}{r!(n-r)!}$$

† To clarify this rule, consider tossing an unbiased die. Since a die has 6 faces, the probability of any single face showing is $\frac{1}{6}$. The probability of a 3 showing, for example, is $\frac{1}{6}$. However, the probability of *either* a 3 or a 5 showing is $\frac{1}{6} + \frac{1}{6} = \frac{1}{3}$. Similarly, the probability of an even-numbered face showing is $\frac{1}{6} + \frac{1}{6} + \frac{1}{6} = \frac{1}{2}$, since this outcome can occur in three distinct ways.

‡ Identical pegs are assumed to be indistinguishable from one another. However, if the pegs were each distinguishable, or unique in some way, then we would be dealing with *permutations*. To clarify this difference, consider placing just two pegs in three holes. The figures below show the three different combinations: $(X)\ (X)\ (\ \ )$; $(X)\ (\ \ )\ (X)$; and $(\ \ )\ (X)$ (X). If the pegs are distinguishable, however, there are six ways; $(X)\ (Y)\ (\ \ )$; $(Y)\ (X)\ (\ \ )$; $(X)\ (\ \ )\ (Y)$; $(Y)\ (\ \ )\ (X)$; $(\ \ )\ (X)\ (Y)$; $(\ \ )\ (Y)\ (X)$.

Note that each of the terms is followed by a factorial operator.   For the case at hand, this formula yields

$$_{20}C_{18} = \frac{20!}{18!2!} = \frac{20 \cdot 19 \cdot 18 \cdots 1}{18 \cdot 17 \cdot 16 \cdots 1 \cdot 2 \cdot 1} = \frac{20 \cdot 19}{2 \cdot 1} = 190$$

Thus, there are 190 different ways in which only two students could guess incorrectly.   Then, $P(18) = 190(.25)^{18}(.75)^2$.

$P(17)$ can be found by a similar process.   The probability of any specific order of 17 corrects and 3 incorrects is $(.25)^{17}(.75)^3$.   There are $_{20}C_{17}$, or 1,140, different possible ways that 17 students could guess correctly.   Hence,

$$P(17) = 1,140(.25)^{17}(.75)^3$$

The student should confirm that $P(16) = 4,845(.25)^{16}(.75)^4$.   This process may be continued until $P(8)$ is found.   In general, the probability for X successes in 20 events would be $P(X) = {}_{20}C_X(.25)^X(.75)^{20-X}$.   Then, $P(X \geq 8)$ would be the sum of these 13 probabilities.   For the record, this probability works out to be approximately .10.   That is, if this experiment were repeated for 100 different samples, we would expect 10 of these samples to show 8 or more correct guesses, on the assumption that the null hypothesis is true.

Knowing the probability of the observed and more extreme outcomes places the researcher in a position of having to decide on the tenability of his assumption, or null hypothesis.   For the present example, most researchers would not consider the outcome extreme enough to suggest rejection of H_0. However, the process of deciding whether or not to reject the null hypothesis is subjective.   Conventionally, outcomes are not considered unusual or extreme unless their probability is less than .05.   If the probability of the given plus all more extreme outcomes is less than .05, the result is described as *significant*.†
For the example, nine or more correct guesses would yield a result significant at the .05 level.   That is, for nine or more correct guesses, the assumption of random guessing by the students could be rejected and the results taken as evidence for ESP ability.   When the researcher decides to reject an H_0 because the data represent an extreme outcome, he is taking a risk of committing a Type I error.   That is, the null hypothesis that he rejects may, in fact, be true for the population.   The *probability* of committing a Type I error is equal to the probability of the outcome upon which the researcher is basing his rejection of the null hypothesis.   For example, if the probability of an outcome is computed and found to be .03, a researcher who rejects the H_0 in this situation would have a probability of .03 of making an incorrect decision.   This is so because in 3 cases out of a 100 when H_0 is *true*, an outcome as extreme as or more extreme than the observed outcome will occur.   This is, in fact, the very meaning of say-

---

† Many researchers prefer to employ a probability of .01 as a critical value rather than .05.

ing that the observed outcome has a probability of .03. For the ESP outcome, a researcher would take a risk of .10 if he decided to reject H_0. For cases in which an H_0 is rejected at a specified level of significance, such as .05, the *maximum* risk of committing a Type I error is .05, since only outcomes less likely than 5 in a 100 would cause a rejection of H_0. On the other hand, when a researcher decides to accept an H_0, he might be committing a Type II error. There is an inverse relationship between Type I and Type II errors with which the student should be familiar. Assuming that all other factors remain constant, the researcher can reduce his risk of a Type I error by simply choosing to reject the null hypothesis only for very unlikely outcomes. Thus, the Type I error could be made negligible by setting the level of significance at, say, .0001, and then rejecting the null hypothesis only for outcomes more extreme than .0001. Unfortunately, if this is done, the risk of a Type II error *increases*. That is, the researcher becomes more likely to accept a false null hypothesis as he reduces his Type I error. As an extreme position, if a researcher decides to accept *all* null hypotheses regardless of the probability associated with his data, he succeeds in reducing his risk of a Type I error to 0. However, his risk of a Type II error becomes 1. Similarly, by rejecting all null hypotheses, the risk of Type II error becomes 0 because no H_0 would ever be accepted. However, the probability of a Type I error is 1. One major concern in designing research studies is to make the risk of a Type II error as small as possible while Type I risk is set at some conventional level, such as .05. Techniques that result in a reduced Type-II-error risk are described as "increasing the *power*, or *sensitivity*, of a statistical test." That is, power (or sensitivity) is the degree to which a statistical test can detect and reject null hypotheses which are, in fact, false. Obviously, a more powerful test is preferred to a less powerful test. The most general procedure for increasing the power of a statistical test is to increase the size of the random sample utilized in the research. Departures from an H_0 are more likely to be detected with large than with small samples.

It should be noted that the use of .05 as a critical value for rejecting a null hypothesis is purely a convention. We do not recommend one's becoming overly attached to this convention. If research of an exploratory or pilot nature is undertaken in a new area of study, an outcome with probability of .10 or even larger may be reason for encouragement that, with more refined techniques, the null hypothesis may prove to be untenable. In view of this, some researchers present actual probability values as well as critical values and decisions concerning the significance or lack of significance of the results.

An additional example of the binomial distribution will be considered in order further to clarify and develop the concepts just presented. Suppose that a prison warden has evidence over a number of years that his present prisoner rehabilitation program results in a 50 percent recidivism rate during the first 6 months among first offenders who are released from prison. Further, assume that he wishes to try out a new rehabilitation program, which has been described in a penology journal, in order to determine if it results in a recidivism rate significantly different from the present one. For this purpose, he installs the new program in his prison for a 3-month trial. At the end of the 3-month period, he randomly selects 10 first offenders who were released under the new program and then keeps track of them for a 6-month period. At the end of this time, three of these individuals have been convicted of new crimes and incarcerated. From previous experience, the pertinent null hypothesis is $H_0: \pi_A = .50$, where π_A is the assumed probability of recidivism for an individual in the population of prisoners going through the new rehabilitation program. This null hypothesis can be tested by use of a binomial distribution. The expected number of reincarcerated individuals under H_0 is $\pi_A n = (.50)10 = 5$, and the observed number is 3. The more extreme possible outcomes are 2, 1, and 0 reincarcerations. Therefore, it is necessary to compute $P(X \leq 3)$. Then,

$$P(X \leq 3) = P(0) + P(1) + P(2) + P(3) \quad \text{or}$$
$$P(X \leq 3) = (.50)^{10} + {}_{10}C_1(.50)(.50)^9$$
$$+ {}_{10}C_2(.50)^2(.50)^8 + {}_{10}C_3(.50)^3(.50)^7$$
$$= \frac{1}{1,024} + \frac{10}{1,024} + \frac{45}{1,024} + \frac{120}{1,1024} = \frac{176}{1,024} = .172$$

There are about 17 chances in 100 of getting 3 or fewer reincarcerated individuals out of 10 under a program whose true recidivism rate is .50. A somewhat different interpretation of this outcome is possible if we consider *deviations* from the expected number of reincarcerations. The prison warden was interested in the possibility not only that the new program might have a lower recidivism rate than .50 but also that its recidivism rate might be larger than .50. When deviations in both directions from expectation are of interest to the researcher, he must use a *nondirectional statistical test*. In the case of the ESP research previously presented, the researcher was involved in a *directional statistical test*, since only deviations that resulted in outcomes above the expectation were of interest. For a nondirectional test in the case of the recidivism study, it is necessary to compute the probability of a deviation of two units (i.e., $|3 - 5| = 2$) from the expected number of reincarcerations. If seven reincarcerations had been observed, this would also represent a *deviation* of two units from expectation. Thus, in order to find the probability under H_0 of observing a deviation as large as or larger than two units, it is necessary to find the sum of $P(X \leq 3)$ and $P(X \geq 7)$. That is, $P(7 \leq X \leq 3) = P(X \leq 3) + P(X \geq 7)$. The student should confirm that $P(X \geq 7) = .172$ for this example. Thus,

$$P(7 \leq X \leq 3) = .172 + .172 = .344$$

This probability represents a nondirectional probability, or the probability of the observed and all equal or more extreme *deviations* from expectation. In this example, the prison warden would conclude that the new program does not differ significantly in recidivism rate from his old program.

The use of directional or nondirectional statistical tests depends upon the research problem. The researcher must decide whether he is interested in specific deviations in one direction only or if outcomes in both directions are important to him. It is quite typical in behavioral science research for the non-directional test to be the relevant one.

Let us now summarize the essential features of a binomial distribution and statistical tests based on it.

1) There exists a dichotomized variable with its two values denoted A and B.

2) The probability of A occurring is $\pi_A = N_A/N$; the probability of B occurring is $\pi_B = N_B/N$. Also, $\pi_A + \pi_B = 1$.

3) π_A and π_B are generally unknown, but some reasonable hypothesis can be made concerning their values. That is, $H_0: \pi_A = \epsilon$, where ϵ is the hypothesized probability of A.

4) A random sample of size n is drawn and the *number* of occurrences of the value A in the sample is r.

5) (a) If a directional test is called for, and if outcomes equal to or *larger* than r are appropriate, then compute $P(X \geq r) = P(r) + P(r + 1) + \cdots + P(n)$ where

$$P(X) = {}_nC_X (\pi_A{}^X)(\pi_B{}^{n-X})$$

and

$${}_nC_X = \frac{n!}{(n - X)!X!}$$

If $P(X \geq r)$ is small (i.e., less than .05), reject the null hypothesis; otherwise accept the null hypothesis. If outcomes equal to or *smaller* than r are appropriate, compute $P(X \leq r) = P(r) + P(r - 1) + \cdots + P(0)$ and reject the null hypothesis only if $P(X \leq r)$ is small.

(b) If a nondirectional test is called for and if r is *larger* than $\pi_A n$ (i.e., if r is larger than the number of outcomes expected by chance), then compute $P(r \leq X \leq 2\pi_A n - r) = P(r \leq X) + P(X \leq 2\pi_A n - r)$. If this probability is small (i.e., less than .05), reject the null hypothesis; otherwise, accept the null hypothesis. If r is *smaller* than $\pi_A n$ (i.e., if r is smaller than the chance number of outcomes), then compute $P(r \geq X \geq 2\pi_A n - r)$ and reject the null hypothesis only if this probability is small.

Chi-square goodness-of-fit test

For large samples, the use of a binomial distribution to test hypotheses concerning categorical data is computationally laborious. Also, if a population is categorical but divided into *more than* two categories, then tests must be based upon a *multinomial distribution*. A multinomial population occurs when a population consists of k mutually exclusive and exhaustive categories. Each category will have an assigned probability and these k probabilities must sum to 1. Thus, a binomial population is a special case of the multinomial, with $k = 2$. Needless to say, computation with a multinomial distribution is even less convenient than with a binomial distribution.

An approximation to probabilities for either binomial or, more generally, multinomial distributions can be obtained by use of a chi-square distribution. Chi-square distributions are continuous, but under a wide range of conditions they provide excellent substitutes for the binomial or multinomial distributions. The major limiting factor for the accuracy of this approximation is the size of the sample. These factors will be discussed in some detail after the use of chi-square is explained. The form of the data and the null hypothesis for a chi-square test are identical with those for a binomial (or multinomial) test. The difference is that the chi-square test utilizes an approximate procedure for computing probabilities. In general, this procedure is to: (1) compute *expected* frequencies for each category assuming that the null hypothesis is true; (2) compute a statistic χ^2 involving discrepancies between observed and expected frequencies; and (3) consult a table of χ^2 values to find the approximate *nondirectional* probability of the observed frequencies under the null hypothesis (Table F.4).

Approximation for binomial data

For the binomial case, the statistic χ^2 has the definitional formula

$$\chi^2 = \sum_{i=1}^{2} \frac{(f_i - f_i')^2}{f_i'}$$

where f_i is the *observed* frequency for the ith category and f_i' is the corresponding *expected* frequency. Note that it is the difference, $f_i - f_i'$, that results in χ^2 being either a large or a small value. Thus, if the observed and expected frequencies differ by small amounts, χ^2 will be a small number. Small values of χ^2 are associated with *large* probabilities of occurrence, which is reasonable if the f_i and f_i' are similar. On the other hand, if the differences, $f_i - f_i'$, tend to be large, χ^2 will also be large. Large values of χ^2 correspond to *small* probabilities of occurrence—as they obviously should.

Let us now consider the use of the chi-square approximation to the binomial distribution. The ESP example previously analyzed by the binomial distribution can now be dealt with by chi-square. Table 2.7 summarizes the

TABLE 2.7   Data from an ESP study

i	Result of guessing suit of card	Observed frequency	Notation
1	Correct	8	f_1
2	Incorrect	12	f_2
	Total	20	n

ESP data in the form of a one-way classification table. The notation column has been shown to clarify the usage adopted for chi-square tests. The null hypothesis concerning the proportion of corrects is $H_0: \pi_A = .25$. Assuming that H_0 is true, we would expect .25, or 25 percent, of the sample to guess correctly and .75, or 75 percent, to guess incorrectly. Since 25 percent of 20 is 5, the *expected* frequency for correct guesses is 5. Similarly, the expected frequency for incorrect guesses is 15. These results are shown by adding an expected frequency column to the one-way classification table (Table 2.8). Note that

TABLE 2.8   Data from an ESP study

i	Result of guessing suit of card	Observed frequency, f_i	Expected frequency, f_i'
1	Correct	8	5
2	Incorrect	12	15
	Total	20	20

the sum of the expected frequencies must equal the sample size n. That is, we must expect a total of n choices because we actually observed n choices. The statistic χ^2 can now be computed. Using the values from Table 2.7 and substituting, we have

$$\chi^2 = \frac{(8-5)^2}{5} + \frac{(12-15)^2}{15} = \frac{9}{5} + \frac{9}{15} = 1.8 + .6 = 2.4$$

Before Table F.4 can be used to determine the probability associated with χ^2, one additional factor must be considered. Table F.4 is set up so that probabilities are read from the top row of the table. Only certain probability values are shown in order to reduce the size of the table. Furthermore, the rows of the table are labeled in terms of *degrees of freedom*. The consideration of degrees of freedom is necessary because there are many different chi-square distributions. Whereas the proof is beyond the scope of this book, it can be shown that the appropriate distribution for a one-way classification table with k rows is the theoretical chi-square distribution with $k - m$ degrees of freedom, where m is the number of different, independent restrictions, in addition to H_0, placed on the expected frequencies. In finding the expected frequencies for the ESP data, we required that

$$\sum_{i=1}^{k} f'_1 = n$$

in addition to $H_0: \pi_A = .25$. Thus, there was one restriction placed on the expected frequencies and $m = 1$. Since $k = 2$, the degrees of freedom for the chi-square test are $k - m = 2 - 1 = 1$. That is, the 1-degree-of-freedom row of Table F.4 is appropriate for our use. In fact, whenever we use the chi-square test for *binomial* data, the degrees of freedom will be 1. An interpretation of degrees of freedom is useful for further understanding of this notion. In a table with two rows, we are free to fill in either one of the two expected frequencies, but then the other one is immediately determined, since the expected frequencies must sum to n. Thus, if n is 20 and if the expected frequency for one outcome is 5, then the expected frequency for the other outcome must be 15, since $5 + 15 = 20$. We were, then, "free" to fill in one cell of the two.

Let us now find the approximate nondirectional probability for the ESP outcome. Looking across the 1-degree-of-freedom row of Table F.4, we find that the computed value, 2.4, does not even approach the critical value (3.84) needed to reject the null hypothesis at the .05 level of significance. Hence, the nondirectional probability of the given outcome plus all more extreme outcomes is greater than .05. From more extensive tables of chi-square, this probability is found to be between .10 and .20. Therefore, we accept the null hypothesis and conclude that we may reasonably assume that the students guessed randomly at the suit of the card. This result is consistent with that from the binomial test. †

The accuracy of the chi-square approximation can be improved by use of a *correction term* in the formula for χ^2. This correction term is equivalent to reducing each of the differences between observed and expected frequencies by .5.

---

† The directional binomial probability was approximately .10; the nondirectional probability is approximately .19.

The correction term can be incorporated into the definitional formula for x^2 as follows:

$$x^2 = \sum_{i=1}^{2} \frac{(|f_i - f_i'| - .5)^2}{f_i'}$$

If the corrected value of x^2 is now computed from ESP data, we obtain

$$x^2 = \frac{(|8 - 5| - .5)^2}{5} + \frac{(|12 - 15| - .5)^2}{15} = 1.25 + .42 = 1.67$$

From more complete tables of chi-square, it is found that the probability associated with $x^2 = 1.67$ is approximately .20. Thus, the use of the correction factor has somewhat improved the approximation of the probability obtained by use of the chi-square. It is recommended that the corrected x^2 always be used instead of the uncorrected x^2 when approximations to binomial probabilities are desired.

Before turning to the use of chi-square tests for multinomial data, we will illustrate the chi-square test for the prison-warden's data considered in the previous subsection. Table 2.9 summarizes these results in a one-way classi-

TABLE 2.9   *Data from prison-warden research*

Outcome	Observed frequency	Expected frequency
Returned to prison	3	5
Did not return to prison	7	5
Total	10	10

fication table. The null hypothesis is $H_0: \pi_A = .50$. Hence, the expected number of incarcerated prisoners is 5 and the expected number of prisoners not returning to prison is also 5. These data have 1 degree of freedom, since the only restriction, in addition to H_0, is that

$$\sum_{i=1}^{2} f_i' = 10$$

Using the corrected formula for x^2, we obtain

$$x^2 = \frac{(|3 - 5| - .5)^2}{5} + \frac{(|7 - 5| - 5)^2}{5} = \frac{1.5^2}{5} + \frac{1.5^2}{5} = \frac{4.5}{5} = .9$$

From the 1-degree-of-freedom row of Table F.4, we find that the nondirectional probability of $\chi^2 = .9$ is greater than .05. This is consistent with the result of .344 from the binomial test.†

Certain limitations should be considered when applying chi-square tests to binomial data. Since chi-square is an approximate test, its overall accuracy is satisfactory only for relatively large sample sizes. If the corrected χ^2 statistic is used for small samples, the probability associated with the χ^2 value will tend to be larger than the corresponding exact probability from the binomial test. In effect, then, the researcher will accept more null hypotheses with the chi-square test than he should according to the exact binomial test. This results in a loss of power for the test, since the Type II error is increased. In some research applications, a researcher may be more concerned about Type I errors than Type II errors. If this is the case, he may justifiably use the corrected χ^2 statistic for testing small samples. In general, however, the sacrifice of power is undesirable; and for samples smaller than, say, 40, the binomial test should be utilized in lieu of a chi-square test. Tables of binomial coefficients and of individual binomial probability are available‡ and can be used to reduce the amount of computation required during binomial tests. If $\pi_A = \pi_B = .50$, the accuracy of the chi-square test based on a corrected χ^2 value is excellent even for very small samples. Thus, the application of a chi-square test to the data in Table 2.9 is appropriate, since $\pi_A = .50$. For samples smaller than 40, this liberalization of the use of chi-square tests is only appropriate when the expected proportions are exactly .50. That is, generally speaking, do not use a chi-square test when n is less than 40 even though π_A is nearly .50.

Approximation for multinomial data

When a population consists of *more than two* discrete, nominal categories, a random sample from that population can be classified into a one-way table and a chi-square test can be used. It is necessary, of course, that pertinent assumptions concerning the proportions of each category in the population be available in order to set up a null hypothesis and to determine expected frequencies. If these requirements are met, the χ^2 statistic can be computed from the general chi-square goodness-of-fit formula

$$\chi^2 = \sum_{i=1}^{k} \frac{(f_i - f_i')^2}{f_i'}$$

† A more complete table of χ^2 values shows that the probability for $\chi^2 \geq .9$ is approximately .342.

‡ See National Bureau of Standards [1950] for samples as large as 49, and Romig [1953] for samples between 50 and 100.

or its computational equivalent:

$$\chi^2 = \sum_{i=1}^{k} \left(\frac{f_i^2}{f_i'}\right) - n$$

where

$$n = \sum_{i=1}^{k} f_i = \sum_{i=1}^{k} f_i'$$

Note that these formulas differ from those for the binomial case only in that the summation is over k categories rather than over two categories. The degrees of freedom for the test will generally be $k - 1$, since the only restriction is

$$\sum_{i=1}^{k} f_i' = n$$

However, in special cases, additional restrictions may be necessary and the degrees of freedom adjusted accordingly.

TABLE 2.10   Data from
rolling a single die

Die face	Observed frequency
1	20
2	12
3	14
4	27
5	15
6	12
Total	100

Consider a population that is defined in terms of the possible different outcomes of rolling a single die. The categories in this population consist of the six outcomes corresponding to the six different faces of a die. If a single die were randomly rolled a very large number of times, and if the die were unbiased, we would expect each face of the die to show on $\frac{1}{6}$ of the total number of outcomes. Let us say that an actual die is rolled 100 times by a mechanical device that produces, in effect, random outcomes and that the observations summarized in Table 2.10 are obtained. If we wish to test these data in order to decide whether this die is unbiased, we can set up the pertinent null hypothesis $H_0: \pi_1 = \pi_2 = \pi_3 = \pi_4 = \pi_5 = \pi_6 = \frac{1}{6}$, where π_1 represents the probability of face 1 showing on a single roll, π_2 represents the probability of face 2 showing on a single roll, and so forth. The expected frequencies in Table 2.11 can be derived from H_0. Each f_i' value is $100(\frac{1}{6}) = 16.67$, since each outcome is equally likely. The sum of the expected frequencies differs from 100 only by a rounding error of .02. χ^2 can now be computed by use of the computational formula.† It is more convenient to perform these computations by steps shown

TABLE 2.11   Observed and expected frequencies for rolling a single die

Die face	Observed frequency	Expected frequency
1	20	16.67
2	12	16.67
3	14	16.67
4	27	16.67
5	15	16.67
6	12	16.67
Total	100	100.02

in columns appended to the one-way classification table (Table 2.12). The sum of the column labeled f_i^2/f_i' is the left-hand term

$$\sum_{i=1}^{k} \frac{f_i^2}{f_i'}$$

from the computational formula for χ^2. Thus, $\chi^2 = 110.28 - 100 = 10.28$. The degrees of freedom for the test are $k - 1 = 5$, since only the restriction

$$\sum_{i=1}^{6} f_i' = 100$$

was placed on the expected frequencies. From Table F.4, use of the 5-degrees-of-freedom row indicates that the nondirectional probability for the observations is greater than .05. This does not provide sufficient evidence to reject H_0 at a conventional level and we must thus conclude that the assumption of an unbiased die was a reasonable one.

† The use of a correction term is not recommended for approximations to multinomial distributions. For a further discussion of this point, see Knetz [1963] and Cochran [1952].

The chi-square goodness-of-fit test can be used in a wide variety of situations. The only general requirements are that the data be represented in two or more mutually exclusive and exhaustive categories and that some reasonable method for determining expected frequencies is available. In summary, the steps are: (1) display the data in the form of a k-rowed one-way classification table; (2) set up the null hypothesis and from it derive the expected frequencies for each of the k rows of the table; (3) compute χ^2 (use corrected χ^2 for 2-rowed tables); (4) determine the degrees of freedom for the test by subtracting the number of independent restrictions imposed on the expected frequencies from the number of rows, k, in the table; (5) use Table F.4 to find the approximate,

TABLE 2.12   Computational columns
for chi-square test

Die face	Observed frequency	Expected frequency	f_i^2	f_i^2/f_i'
1	20	16.67	400	24.00
2	12	16.67	144	8.64
3	14	16.67	196	11.76
4	27	16.67	729	43.74
5	15	16.67	225	13.50
6	12	16.67	144	8.64
Total	100	100.02		110.28

nondirectional probability of occurrence for the observed frequencies; (6) make a decision concerning the acceptance or rejection of the null hypothesis.

For those who wish to study a more complex example of the chi-square goodness-of-fit test, Technical Appendix B presents a case in which three restrictions are imposed on the expected frequencies prior to the computation of χ^2.

Problems

2.1. A parent went to a pet shop to select either a kitten or a puppy for his child. The shop had one black and three white puppies, and three black and three white kittens. If the parent were to make a selection at random, what is the probability that the pet selected is either a puppy or is a black animal?

2.2. A bag contains 10 blue balls and 20 red balls. In two random draws from the bag, what is the probability of drawing exactly one blue ball and one red ball in selecting (*a*) with replacement and (*b*) without replacement of the first ball?

2.3. A member of a garden club orders nine special bulbs from an exporter in Holland. The purchaser requests three blue, three red, and three white bulbs. Upon delivery, the woman discovers that the bulbs are mixed together and indistinguishable from one another with respect to color. If the bulbs are divided into three groups of three bulbs, what is the probability that when the groups bloom, one will be all blue, one all red, and one all white?

2.4. How many permutations of size 3 can be taken from the letters M, N, O, P, Q, and R?

2.5. Use the binomial formula to determine the probability of getting exactly three heads in eight tosses of a fair coin.

2.6. What are the different types of tabular procedures that can be used with univariate, categorical data?

2.7. What reasons are there for using derived measures rather than measures in their original form?

2.8. What graphical procedures are appropriate to univariate, categorical data?

2.9. Under what conditions can a sample be appropriately called "random"?

2.10. What is stratified random sampling and how does this technique differ from simple random sampling?

2.11. The major areas of students in an education course were special education, 5; administration, 7; elementary education, 15; secondary education, 9; counseling, 11; and library science, 3. Make a table showing the frequencies and percentages of students falling into each group. Illustrate this information in the form of (*a*) a bar diagram and (*b*) a pie diagram.

2.12. What effect does the correction factor in the chi-square statistic as applied to dichotomous data have upon the magnitude of chi-square and upon Type I and Type II errors?

2.13. What are two general requirements for the use of chi-square goodness-of-fit tests?

2.14. If the difference between observed and expected frequencies is small, chi-square will be (large; small) and the probability of occurrence will be (large; small). If the difference is large, chi-square will be (large; small) and the associated probability will be (large; small).

2.15. In a random sample of 10 babies, 9 are boys and 1 is a girl. Assuming that boy and girl babies are equally likely, calculate chi-square (*a*) using the correction factor and (*b*) not using the correction factor.

2.16. A coin was tossed 200 times resulting in 115 heads and 85 tails. Use the chi-square statistic to test the hypothesis that the coin is fair. Use the

.05 level of significance and apply the chi-square formula (*a*) with the correction factor and (*b*) without the correction factor.

2.17. For a given examination, there were expected to be 5 percent A's, 20 percent B's, 40 percent C's, 25 percent D's, and 10 percent F's. The actual distribution of grades for a class of 45 students was 2 A's, 10 B's, 15 C's, 15 D's, and 3 F's. Using a test at the .05 level of significance, decide if the observed distribution of grades is consistent with expectation.

Three

descriptive and inferential methods for bivariate categorical data

3.1 Descriptive techniques

Tabular techniques

When the purpose of a research investigation is to study the relationship between two different variables, the data will be bivariate. A bivariate distribution has a value on *each* variable for *each* individual in the population, and the simplest such distribution occurs when each of the variables is dichotomous. To illustrate this case, Table 3.1 displays data for both sex and handedness of a population of 10 students. The relationship between sex and handedness for these 10 students can be shown more clearly in a two-way classification, or *contingency*, table (Table 3.2). The numbers appearing in the cells of the contingency table represent frequencies of occurrence for specific pairs of categories. Thus, the upper-left cell contains four students, each of whom are male and right-handed. Similarly, the lower-right cell contains the one left-handed female in the population. The sums of the *columns* of the table are identical with the distribution of sex alone, whereas the sums of *rows* of the table represent the distribution of handedness. These row and column sums are referred to as *marginal frequencies* because they appear in the margins of the contingency table. On the other hand, the frequencies appearing in the cells of the contingency table are called *conditional frequencies* because they represent the frequencies for one variable under the condition that the second variable has a certain value. For example, the frequency of 2 for left-handed males may be interpreted as the

TABLE 3.1 *Bivariate distribution of sex and handedness for 10 students*

Student number	Sex	Handedness	Student number	Sex	Handedness
1	Male	Right	6	Female	Right
2	Male	Right	7	Female	Right
3	Female	Left	8	Male	Left
4	Male	Left	9	Male	Right
5	Female	Right	10	Male	Right

TABLE 3.2 *Contingency table for sex and handedness data*

		Sex		
		Male	Female	Total
Handedness	Right	4	3	7
	Left	2	1	3
	Total	6	4	10

number of males in the population *on the condition that* they are left-handed (or, equivalently, the number of left-handed students in the population *on the condition that* they are male).

Contingency tables may be constructed for cases in which one or both of the measures take on more than two values. For these cases, the contingency table will contain as many cells as the product of the numbers of values for the two variables. Thus, a contingency table for two dichotomous variables contains four cells (i.e., $2 \cdot 2 = 4$ cells, since each variable has two values), and a contingency table containing 20 cells would be required for the case in which one variable takes on 5 values and the second variable takes on 4 values.   Table 3.3

TABLE 3.3   *Marital status for males in four different social clubs*

	\multicolumn{4}{c}{Social club}	
	1	2	3	4	Total
Married	133	164	155	106	558
Single	36	57	40	37	170
Total	169	221	195	143	728

illustrates the case of an eight-celled contingency table where one variable, marital status, is dichotomous, whereas the second variable, social club, has four values.   The cell values from this table represent specific combinations of marital status and membership in a social club.   Thus, social club 2 contained 164 married and 57 single males, and so forth for the remaining columns of the table.

Derived measures

As in the case of one-way classification tables, the cell and marginal frequencies from a contingency table may be converted to percentages or proportions. Changing row and column sums to proportions is a straightforward procedure. To illustrate this process, let us consider the sex/handedness data previously discussed (Table 3.2).   When each marginal total is converted to a proportion, this result appears as in Table 3.4.   In general, if the total frequency of cases in

TABLE 3.4   *Marginal proportions*

		\multicolumn{2}{c}{Sex}	
		Male	Female	Total
Handedness	Right			.70
	Left			.30
	Total	.60	.40	1.00

a contingency table is N, then the marginal frequencies are transformed to proportions by dividing each marginal sum by N (for example, $.70 = 7/10$ and $.30 = 3/10$). Note that both the row distribution and column distribution of marginal proportions must sum to 1.00.

The task of changing cell, or conditional, frequencies to proportions is complicated by the fact that there are three different ways of carrying this out and each yields unique results. These methods are: (1) Conversion by rows, that is, each row of the table is changed to proportions by dividing the conditional frequencies of each row by the marginal sum for that row (Table 3.5).

TABLE 3.5   *Conversion by rows to proportions—sex/handedness data*

		Male	Female	Total
		Sex		
Handedness	Right	.57	.43	1.00
	Left	.67	.33	1.00
	Total	.60	.40	1.00

Note that the proportions will sum to 1.00 for each row of the table. For this example, conversion by rows allows statements such as "57 percent of the right-handers are male" or "67 percent of the left-handers are male." If percentage or proportionate statements concerning rows of the contingency table are desired, then conversion by rows is the appropriate procedure. (2) Conversion by columns, that is, each column of the table is changed to proportions by dividing the conditional frequencies of each column by the marginal sum for that column (Table 3.6). In this case, each column will sum to 1.00, and this pro-

TABLE 3.6   *Conversion by columns to proportions—sex/handedness data*

		Male	Female	Total
		Sex		
Handedness	Right	.67	.75	.70
	Left	.33	.25	.30
	Total	1.00	1.00	1.00

cedure is appropriate when statements concerning columns are desired (i.e., "75 percent of the females are right-handed" or "33 percent of the males are left-handed").   (3) Conversion by total N, that is, each conditional frequency is divided by N (Table 3.7).   For this case, the sum of *all* cells will be 1.00.   Also,

TABLE 3.7   *Conversion by total N to proportions—sex/handedness data*

		Male	Female	Total
		Sex		
Handedness	Right	.40	.30	.70
	Left	.20	.10	.30
	Total	.60	.40	1.00

note that when row and column sums are formed, they agree with the values obtained in Table 3.4 for conversion of marginal frequencies to proportions. This procedure should be used whenever statements concerning the occurrence of specific combinations of rows and columns are desired.   For example, from Table 3.7 it may be concluded that 40 percent of the total population consists of right-handed males and 10 percent of left-handed females.

Conversion to proportions by either rows or columns is the more common technique, since it allows some important comparisons not possible when conversion is based on the total N.   For example, the proportions derived from the columns (Table 3.6) allow the observation that 75 percent of the females and only 67 percent of the males are right-handed.   Thus, right-handedness occurs in greater proportion among females than among males for this population. The proportions based on total N (Table 3.7) obscure this fact and tell us only that of the total population 40 percent are right-handed males and 30 percent are right-handed females.

By studying proportions derived either by row or column conversion, it is possible to draw conclusions concerning relationships between the row and column variables.   Referring once again to the sex/handedness data (Table 3.6), we see that the following language may be used in discussing the higher proportion of right-handedness among females: "There is an *association* between handedness and sex of individuals in this population; this association operates so that females tend to show a larger proportion of right-handedness than do males, whereas males tend to show more left-handedness than do females." Another way of expressing this relationship is to state that one variable is *dependent* upon the other. Thus, for this example, handedness is dependent upon sex. Yet another way of expressing this relationship is to say that sex and handedness are *correlated* in this population. †

To clarify what is meant by an association between two variables, let us consider a new example in detail. Suppose that a survey was made of ninth graders in a large school. Each student was asked if he had attended a social function (e.g., dance, party) on the previous Friday night. Furthermore, assume that the school enrolled 220 boys and 280 girls or a total of 500 children. On the Friday night for which the social behavior was surveyed, 200 of these children attended a social function and 300 did not. These facts can now be summarized in the form of a 2 × 2 contingency table (Table 3.8), although

TABLE 3.8   *Social behavior of 500 students*

	Sex		
	Boys	*Girls*	*Total*
Attended			200
Did not attend			300
Total	220	280	500

from the information given it is not yet possible to fill in the cell frequencies. At this point, let us speculate on the *possible* relationships which may exist in these data. Consider first the case in which there is no *association* between sex and whether or not students attended social functions. For this case, we would expect boys and girls to attend social functions in equal proportions. Since 200/500, or 40 percent, of the total student body attended social functions, this means that 40 percent of the boys and 40 percent of the girls would attend social functions. Table 3.9 shows conditional frequencies and conditional propor-

† Statisticians utilize at least four different, but essentially equivalent, ways of talking about a relationship between two variables. If X and Y are the two variables, then one may say any one of the following: X is related to Y; X is associated with Y; X is dependent upon Y; or X is correlated with Y. Although these four ways of describing relationships are typically used under slightly different circumstances, they may be viewed as essentially the same in meaning.

TABLE 3.9   *Social behavior of 500 students*

	Sex		
	Boys	Girls	Total
Attended	88 (.40)	112 (.40)	200 (.40)
Did not attend	132 (.60)	168 (.60)	300 (.60)
Total	220 (1.00)	280 (1.00)	500 (1.00)

tions (in parentheses) for the case of no association.† A useful way of viewing the results in Table 3.9 is to consider how well social activity could have been *predicted* if only the sex of the student were known.   For the case of no association, knowing the sex of the student gives no clue about whether or not the student attended a social function, since boys and girls attended social functions in equal proportions.   On the other hand, if sex and social activity were correlated, this fact could be used to make predictions.   Consider an extreme case of association where only girls attended social functions.   These results are shown in Table 3.10.   The strong association between sex and social activity is indicated

TABLE 3.10   *Social behavior of 500 students*

	Sex		
	Boys	Girls	Total
Attended	0 (.00)	200 (.71)	200 (.40)
Did not attend	220 (1.00)	80 (.29)	300 (.60)
Total	220 (1.00)	280 (1.00)	500 (1.00)

† The example shows conversions to proportions by columns.   The same result would be obtained by converting rows, however.   Thus 220/500, or 44 percent, of the total population were boys.   Then we would expect 44 percent of those who attended social functions to be boys and 44 percent of those who did not attend social functions to be boys.   If these results are represented in the contingency table, we obtain the following table.

	Boys	Girls	Total
Attended	88 (.44)	112 (.56)	200 (1.00)
Did not attend	132 (.44)	168 (.56)	300 (1.00)
Total	220 (.44)	280 (.56)	500 (1.00)

by the proportions of boys and girls who attended social functions. Whereas 71 percent of the girls attended a social function, none of the boys did. For this case, prediction of social activity is clearly aided by knowledge of the sex of the individual. If a student is a boy, then we can predict with certainty that he did not attend a social function. For girls, the prediction that she did attend a social function would be correct 71 times in a 100.

In actual research studies, it is more typical for relationships to fall somewhere between the extremes discussed above. An example of an intermediate relationship is presented in Table 3.11. Note that this example has been con-

TABLE 3.11   *Social behavior of 500 students*

	Sex		
	Boys	*Girls*	*Total*
Attended	100 (.45)	100 (.36)	200 (.40)
Did not attend	120 (.55)	180 (.64)	300 (.60)
Total	220 (1.00)	280 (1.00)	500 (1.00)

structed so that a greater proportion of boys than girls attended social functions. When frequencies have been reduced to the form of 2 × 2 contingency tables, there are procedures for computing derived measures that indicate the *degree* or *intensity* of the relationship between the row and column variables. These measures are known as *coefficients of correlation*. There are many different coefficients of correlation, not all of which are appropriate for categorical data. The one coefficient that is generally appropriate for categorical data presented in the form of 2 × 2 contingency tables is known as the *phi coefficient*, and is symbolized as Φ. We will first consider the mathematical procedures required to compute the value of Φ from data in a 2 × 2 contingency table and then the problem of interpreting the meaning of Φ.

In order to write a formula for Φ, a standard notation is set up for the cell and marginal frequencies of a 2 × 2 contingency table (Table 3.12). The symbol f, referring to frequency, is given *two* subscripts. The first subscript stands for the *row* of the table and the second for the *column*. Thus, f_{12} is the observed frequency in the cell at row 1, column 2. Also, marginal frequencies have a special notation. Thus, $f_{1.}$ is the sum for the first row (that is, $f_{11} + f_{12} = f_{1.}$). Note that the dot is used to replace the subscript over which

TABLE 3.12   Standard notation for observed frequencies in 2 × 2 tables

		Variable 2		
		1	*2*	*Total*
Variable 1	1	f_{11}	f_{12}	$f_{1.}$
	2	f_{21}	f_{22}	$f_{2.}$
	Total	$f_{.1}$	$f_{.2}$	$f_{..}$

the marginal frequency has been summed. Also, the total number of cases is $f_{1.} + f_{2.} = f_{.1} + f_{.2} = f_{..}$. The phi coefficient is then computed as

$$\Phi = \frac{|f_{11}f_{22} - f_{12}f_{21}|}{\sqrt{f_{1.}f_{2.}f_{.1}f_{.2}}}$$

Note that the absolute-value operator is applied to the numerator of the formula and that as a result Φ will always be a positive quantity. If Φ is computed for the sex/handedness data previously discussed (Table 3.2), the result is as follows: $f_{11} = 4$, $f_{12} = 3$, $f_{21} = 2$, $f_{22} = 1$; then

$$\Phi = \frac{|4 \cdot 1 - 3 \cdot 2|}{\sqrt{7 \cdot 3 \cdot 6 \cdot 4}} = \frac{2}{\sqrt{504}} = \frac{2}{22.45} = .09$$

In order to understand the meaning of a computed value of Φ, it is necessary to know that Φ has a minimum of 0, which indicates that *no association* exists between the row and column variables. On the other hand, Φ has a maximum value of 1.00, and this value occurs only when the row and column variables are *perfectly* associated or correlated. To illustrate perfect correlation, consider the data in Table 3.13, which shows sex/handedness data for a new population of 30

TABLE 3.13   Sex/handedness data for 30 students

			Sex	
		Male	*Female*	*Total*
Handedness	Right	16	0	16
	Left	0	14	14
	Total	16	14	30

students. In this population, each male is right-handed and each female is left-handed. Thus, handedness can be predicted perfectly from a knowledge of a student's sex. When Φ is computed for these data, we obtain:

$$\Phi = \frac{|16 \cdot 14 - 0 \cdot 0|}{\sqrt{(16 \cdot 14 \cdot 16 \cdot 14)}} = \frac{16 \cdot 14}{16 \cdot 14} = 1.00$$

Values of Φ between 0 and 1 indicate varying degrees of intensity of relationship. The closer to 0 a value is, the lower the intensity of the correlation. To illustrate some possible intermediate values, let us consider the Φ coefficients computed from the following tables, which represent sex/handedness data for three different populations.

Example 1

	Male	Female	
Right	12	8	20
Left	8	12	20
	20	20	40

$$\Phi = \frac{|12 \cdot 12 - 8 \cdot 8|}{\sqrt{(20 \cdot 20 \cdot 20 \cdot 20)}}$$

$$= \frac{80}{400} = .20$$

Example 2

	Male	Female	
Right	15	5	20
Left	5	15	20
	20	20	40

$$\Phi = \frac{|15 \cdot 15 - 5 \cdot 5|}{\sqrt{(20 \cdot 20 \cdot 20 \cdot 20)}}$$

$$= \frac{200}{400} = .50$$

Example 3

	Male	Female	
Right	2	18	20
Left	18	2	20
	20	20	40

$$\Phi = \frac{|2 \cdot 2 - 18 \cdot 18|}{\sqrt{(20 \cdot 20 \cdot 20 \cdot 20)}}$$

$$= \frac{320}{400} = .80$$

Once values of Φ have been computed, it is possible to compare the relationship of two variables in different populations or the relationship of different pairs of variables in the same population. Thus, the three populations illustrated above show increasing relationships between sex and handedness, since Φ progresses from .20 to .80 as one moves from population 1 to population 3. As an exercise, the student should compute conditional proportions in order to confirm the fact that in populations 1 and 2 right-handedness is associated with males and in population 3 the opposite is true.

When data are summarized in contingency tables that are larger than 2×2, it is still possible to study relationships between the row and column variables—although no coefficient of correlation is presented here for this case. Consider the data in Table 3.14, which shows the results of a survey of literary

TABLE 3.14   *Exemplary 4 × 3 contingency table*

	Upper	Middle	Lower	Total
	\multicolumn{3}{c}{*Socioeconomic class*}		
Adventure	10 (.05)	95 (.19)	145 (.48)	250 (.25)
Mystery	50 (.25)	100 (.20)	100 (.33)	250 (.25)
Biography	40 (.20)	180 (.36)	30 (.10)	250 (.25)
Historical	100 (.50)	125 (.25)	25 (.08)	250 (.25)
Total	200 (1.00)	500 (1.00)	300 (.99)	1,000 (1.00)

taste for a community of 1,000 adults. The column variable is socioeconomic class; and the column sums show the distribution for this variable (i.e., in this community, there were 200 upper-class, 500 middle-class, and 300 lower-class adults). The row variable consists of four categories of literature, and the row sums indicate that the preferences of the population were evenly divided among the four types. The cell values can be studied in order to determine what type of relationship exists between socioeconomic class and preferred type of literature. Conditional proportions computed by column are shown in parentheses. It is easy to see that the upper class tends to prefer historical literature, the middle class tends to prefer biography, and the lower class tends to prefer mystery and adventure.

In general, larger-order contingency tables may contain very complex relationships, and these can only be discovered by a careful study of the conditional proportions. It is often useful to compute the proportions first by rows and then, separately, by columns, since some overlooked relationship may be highlighted by this procedure.

3.2   Chi-square test for independence

When a random sample rather than the entire population is available to a researcher, he must utilize techniques of inferential statistics.   For cases involving bivariate data, the chi-square statistic can be used to test whether the sample may be considered to have come from a population in which there is no association or correlation between the two categorical variables—i.e., that the two variables are independent of one another.   Due to sampling variability, a contingency table may seem to show an association between two variables even when these variables are independent in the population.   In this section, we shall consider procedures for testing hypotheses of independence for data resulting from survey research.   The appropriate statistical tests are based on chi-square distributions.   In the next section, it will be seen that the same test procedures are appropriate for experimental studies involving two or more groups and one categorical measured variable.   However, the interpretations of these two situations are somewhat different and are treated separately.

The use and interpretation of the coefficient of correlation for 2 × 2 contingency tables based on populations was introduced earlier in this chapter. The notion of examining tables larger than 2 × 2 for relationships was also explained.   Consider now a situation in which data on two variables from a random sample rather than from an entire population are reduced to the form of a contingency table.   For example, suppose that a study of the social behavior of a new population of ninth graders is undertaken.   Furthermore, assume that in this case complete enumeration of the population is impractical but that a random sample of 150 ninth graders has been selected and sex and social activity have been recorded.   Then, Table 3.15 shows the bivariate distribution

TABLE 3.15   Social behavior of 150 students

	Sex		
	Boys	*Girls*	*Total*
Attended	34	58	92
Did not attend	36	22	58
Total	70	80	150

of sex and social behavior for this sample. We can, of course, convert these frequencies to proportions in order to compare the observed proportions of boys
and girls who attended a social function (Table 3.16).† *In the sample, 72.50*

TABLE 3.16   *Conversion to proportions by columns*

	Boys	Girls	Total
		Sex	
Attended	.4857	.7250	.6133
Did not attend	.5143	.2750	.3867
Total	1.0000	1.0000	1.0000

percent of the girls and 48.57 percent of the boys did attend a social function.
However, we know that the results of random sampling are affected by sampling
variability, and we cannot, at this point, decide that *in the population* a larger proportion of girls than boys attended social functions. First, it is necessary to
determine how likely the observed outcome is under the assumption that boys
and girls did attend social functions in equal proportions in the population.
That is, we can set up the null hypothesis that sex and social behavior are independent of one another. Then, by use of a chi-square statistic, we can find the
approximate‡ probability of outcomes *as extreme as or more extreme in both directions* than the observed outcome. If this probability is sufficiently small (say,
.05 or less), the null hypothesis may be rejected in favor of the assumption that
some real association between sex and social activity exists in the population.
Rejecting the null hypothesis, in other words, is equivalent to deciding that *in
the population* the proportion of girls who attended social functions is different
from the proportion of boys who attended social functions. Of course, as with
all statistical tests, there is the possibility of a Type I error when rejecting a
null hypothesis.

The mechanics for performing a chi-square test of independence are, for
the most part, the same as for a goodness-of-fit test. First, a null hypothesis

---

† The computations are carried to four decimal places so that rounding errors will not be a
serious problem in later computations.

‡ As was true for the goodness-of-fit case, chi-square provides an approximation that is
computationally more convenient than the exact distribution. For the case of two dichotomized variables, the exact distribution is known as a hypergeometric distribution. For small
samples, direct computation of probabilities using a hypergeometric distribution is feasible.
For details on this procedure, see Fisher [1946] or Siegel [1956].

must be set up. Since we are assuming that sex and social activity are independent, the null hypothesis for the example may be written as $H_0\colon \pi_{A_b} = \pi_{A_g}$, where π_{A_b} is the probability that a boy attended a social function and π_{A_g} is the similar probability for a girl (that is, A stands for the event of attending a social function and b and g stand for boy and girl, respectively). It is now necessary to find some reasonable numerical value for π_{A_b} before computing expected frequencies for the cells of the 2×2 contingency table. This problem can be solved from information already contained in the contingency table. Note that the *marginal distribution* for social activity reveals that 61.33 percent of the *total sample* of ninth graders did attend a social function. This is the most reasonable value we have available for estimating the assumed equal proportions π_{A_b} and π_{A_g}. Thus, our common estimate of π_{A_b} and π_{A_g} is .6133. Of course, this also implies that our estimate of π_{B_b} and π_{B_g} is .3867 (i.e., the estimate of those not attending a social function). Expected frequencies, along with the original observed frequencies, for the cells of the table are shown in Table 3.17.† The

TABLE 3.17   *Observed and expected frequencies for 150 students*

	Sex		
	Boys	*Girls*	*Total*
Attended	34/42.93	58/49.07	92
Did not attend	36/27.07	22/30.93	58
Total	70	80	150

expected frequency of boys who attended social functions is simply 61.33 percent of the total number of boys in the sample. That is, 42.93 was computed as the expected frequency for boys attending social functions by multiplying 70 by .6133. Similarly, the expected frequency for girls attending social functions was 80(.6133), or 49.06. Note that the expected frequencies sum, within rounding errors, to the *marginal* totals. Since the total observed number of ninth graders reported attending social functions was 92, we must also *expect* 92 students to attend social functions. Similarly, 70 boys and 80 girls were observed in the sample; hence, we must also *expect* a total of 70 boys and 80 girls. For purposes of notation, the convention adopted for expected frequencies in goodness-of-fit cases is utilized for contingency tables also. Thus, a

† The value *above* the diagonal in a cell is an observed frequency; the value *below* the diagonal is an expected frequency.

prime indicates an expected frequency, and $f'_{11} = 42.93$, $f'_{12} = 49.06$, and so forth. From the observed and expected frequencies, a chi-square statistic can be defined:

$$\chi^2 = \sum_{i=1}^{2} \sum_{j=1}^{2} \frac{(f_{ij} - f'_{ij})^2}{f'_{ij}}$$

The use of a double-summation operator is necessitated by the appearance of two subscripts on the f and f' values. The rule for calculating with a double-summation operator is to start by setting the subscript for the *leftmost* summation operator at its lowest value (i.e., 1 for the present formula) and then substituting the values of the subscript indicated by the rightmost summation operator. Then, reset the subscript for the leftmost summation operator at its second value (that is, 2 for the present formula), and then repeat the process for the rightmost summation operator. In effect, the double-summation operator indicates summing across rows and down columns of the contingency table so that every cell is included in the computation. Following the above rule, the chi-square formula can be written in detail for a 2 × 2 contingency table as

$$\chi^2 = \frac{(f_{11} - f'_{11})^2}{f'_{11}} + \frac{(f_{12} - f'_{12})^2}{f'_{12}} + \frac{(f_{21} - f'_{21})^2}{f'_{21}} + \frac{(f_{22} - f'_{22})^2}{f'_{22}}$$

A correction term should be incorporated into the chi-square formula whenever tests are based on 2 × 2 contingency tables (as was true for the two-celled goodeness-of-fit case). Corrected chi-square is computed from the formula

$$\chi^2 = \sum_{i=1}^{2} \sum_{j=1}^{2} \frac{(|f_{ij} - f'_{ij}| - .5)^2}{f'_{ij}}$$

When the observed and expected frequencies from Table 3.17 are substituted in the formula for corrected chi-square, we obtain

$$\chi^2 = \frac{(|34 - 42.93| - .5)^2}{42.93} + \frac{(|58 - 49.07| - .5)^2}{49.06}$$
$$+ \frac{(|36 - 27.07| - .5)^2}{27.07} + \frac{(|22 - 30.93| - .5)^2}{30.94} = 8.03$$

Before using Table F.4 to determine the probability for a chi-square value of 8.03, it is necessary, once again, to consider the number of degrees of freedom. The same reasoning applied to goodness-of-fit cases can be used to determine the number of degrees of freedom for a contingency table. That is, for a contingency table with a total of kc cells (i.e., with c columns and k rows), the number of degrees of freedom is $kc - m$, where m is the number of independent restrictions, other than the null hypothesis, imposed on the expected frequencies.

To find the expected frequencies shown in Table 3.17, three independent restrictions were imposed:

1) $\displaystyle\sum_{i=1}^{2}\sum_{j=1}^{2} f'_{ij} = \sum_{i=1}^{2}\sum_{j=1}^{2} f_{ij} = 150$

that is, the total sum of the expected frequencies had to be the total sample size, 150.

2) $\displaystyle\sum_{j=1}^{2} f'_{1j} = \sum_{j=1}^{2} f_{1j} = 92$

that is, the sum of the first *row* of the table had to be 92, the observed sum for that row.

3) $\displaystyle\sum_{i=1}^{2} f'_{i1} = \sum_{i=1}^{2} f_{i1} = 70$

that is, the sum of the first *column* had to be 70, the observed sum for that column.

Two additional restrictions were imposed on the expected frequencies, but they are not independent of the three already mentioned. Thus, the restriction

$$\sum_{j=1}^{2} f'_{2j} = \sum_{j=1}^{2} f_{2j} = 58$$

(i.e., the sum of the second row had to be 58) is implied by restrictions 1 and 2 above, since $58 = 150 - 92$. Also, the restriction

$$\sum_{i=1}^{2} f'_{i2} = \sum_{i=1}^{2} f_{i2} = 80$$

(i.e., the sum of the second column had to be 80) is implied by restrictions 1 and 3 above, since $80 = 150 - 70$. Since there are kc, or $2 \cdot 2 = 4$ cells in a 2×2 contingency table and since $m = 3$, the number of degrees of freedom is $4 - 3 = 1$. It should be noted that the number of degrees of freedom for any chi-square test of independence based on a 2×2 contingency table is always 1.

The final step in testing the null hypothesis of independence for sex and dance attendance can now be taken. From the 1-degree-of-freedom row of Table F.4, we find that $\chi^2 = 8.03$ indicates a probability of occurrence of less than .005. This would be taken as evidence for rejecting the null hypothesis at a conventional level (say, .05). We can now state that the test indicates that sex and social activity are not independent in the population. With rejection of H_0, we decide that $\pi_{A_b} \neq \pi_{A_g}$ and that some association exists between sex and social behavior in the population of ninth graders. Inspection of the table of observed proportions (Table 3.16) reveals that this association operates in such a way that girls are more likely than boys to attend a social function; or, conversely, that boys are less likely than girls to attend a social function.

A somewhat different interpretation of the above result can be made if we talk in terms of a *correlation* between sex and social behavior. The phi coefficient for samples can be computed from the table of observed frequencies (Table 3.15). Whereas the computation of the phi coefficient from a sample is identical with the procedure presented for application to population frequencies, it is denoted by the lowercase Greek letter phi ϕ. Because of sampling variability, ϕ may be larger than zero, even though $\Phi = 0$ in the population. Thus, the null hypothesis H_1: $\Phi = 0$ is relevant and can be tested. However, it turns out that testing this null hypothesis is equivalent to the chi-square test of independence just presented. Thus, if we reject the null hypothesis from a chi-square test of independence based on a 2×2 contingency table, we also conclude that $\Phi > 0$. Alternatively, if we accept the independence of row and column variables from a 2×2 contingency table, we also conclude that $\Phi = 0$. Therefore, the results of the chi-square test of independence can be interpreted as indicating either that (1) there is no correlation between the two variables when the null hypothesis is accepted or (2) there is a correlation other than 0 between the two variables when the null hypothesis is rejected.

The general procedures for testing independence by use of a chi-square distribution can be extended to random samples summarized in any contingency table larger than 2×2. In outline, the steps are as follows. (1) Present the data in the form of a $k \times c$ contingency table, where k is the number of rows and c is the number of columns in the table. (2) Set up the relevant null hypothesis. In general, this may be mathematically somewhat complex, but it follows the form

$$\pi_{11} = \pi_{12} = \cdots = \pi_{1c}$$
$$\pi_{21} = \pi_{22} = \cdots = \pi_{2c}$$
$$\cdots \cdots \cdots \cdots \cdots \cdots$$
$$\pi_{k1} = \pi_{k2} = \cdots = \pi_{kc}$$

where the double subscript on the probabilities conforms to the usual contingency-table usage. In most cases, it is more convenient to omit this mathematical statement of H_0 and to write simply that the null hypothesis states that the row and column variables are independent of one another in the population. (3) Compute an expected frequency for each cell of the $k \times c$ table by estimating population proportions from the marginal distributions, as was done for 2×2 tables. (4) Compute

$$\chi^2 = \sum_{i=1}^{k} \sum_{j=1}^{c} \frac{(f_{ij} - f'_{ij})^2}{f'_{ij}}$$

or its computational equivalent

$$\chi^2 = \sum_{i=1}^{k} \sum_{j=1}^{c} \left[\frac{f_{ij}^2}{f_{ij}'} \right] - f..$$

Note that a correction term is *not* used for tables larger than 2 × 2.   (5) Determine the number of degrees of freedom for the chi-square test.   In general, this will be given by the formula $df = (k-1)(c-1)$.   (6) Compare the computed χ^2 with values in the appropriate row of Table F.4 and make a decision concerning the acceptance or rejection of the null hypothesis.

Now let us consider a numerical example illustrating the application of a chi-square test of independence to variables that take on more than two values. A questionnaire is designed by a researcher to evaluate certain aspects of an individual's beliefs concerning religion and morality. The population of interest is all undergraduate students at the college where the researcher is a staff member.   He administers the questionnaire to a random sample of 200 students drawn from this population.   The researcher is especially interested in determining what relationship exists between responses to two specific items on the questionnaire. These items, along with the possible student responses, were:

ITEM *A*   Organized religion is an extremely important force in human life.

( ) Agree   ( ) Neutral   ( ) Disagree

ITEM *B*   Personal morality is independent of the age and culture in which a person lives.

( ) Agree   ( ) Neutral   ( ) Disagree

Analysis of the questionnaire responses was performed and results for items *A* and *B* were shown in the form of a contingency table (Table 3.18).   The inde-

TABLE 3.18   *Observed frequencies for two questionnaire items*

		Item A			
		Agree	*Neutral*	*Disagree*	*Total*
	Agree	25	30	25	80
Item B	Neutral	10	30	10	50
	Disagree	5	30	35	70
	Total	40	90	70	200

pendence of responses to items A and B can be tested by a chi-square test.   The null hypothesis can be written as

$$H_0: \begin{cases} \pi_{11} = \pi_{12} = \pi_{13} \\ \pi_{21} = \pi_{22} = \pi_{23} \\ \pi_{31} = \pi_{32} = \pi_{33} \end{cases}$$

where π_{ij} is the population proportion for the cell at the intersection of the ith row and the jth column of the contingency table.   The numerical estimates of these population proportions can be obtained by dividing each *row* sum by 200. Expected frequencies are obtained by successively multiplying each *column* sum by the estimated population proportions.   Table 3.19 summarizes these

TABLE 3.19   *Observed and expected frequencies for two questionnaire items*

		Item A			
		Agree	*Neutral*	*Disagree*	*Total*
	Agree	25/16.00	30/36.00	25/28.00	80
Item *B*	Neutral	10/10.00	30/22.50	10/17.50	50
	Disagree	5/14.00	30/31.50	35/24.50	70
	Total	40	90	70	200

expected frequencies (below the diagonal in a cell) and the original observed frequencies (above the diagonal in a cell).   By use of the computational formula, χ^2 is found to be

$$\chi^2 = \frac{25^2}{16.00} + \frac{30^2}{36.00} + \cdots + \frac{35^2}{24.50} - 200 = 222.45 - 200 = 22.45$$

The number of degrees of freedom is found by formula to be

$$df = (k-1)(c-1) = (3-1)(3-1) = 4$$

From the 4-degrees-of-freedom row of Table F.4, we find that $\chi^2 = 22.45$ indicates a probability of occurrence less than .005.   Thus, we reject the null hypothesis and conclude that some association exists in the population between responses on the two questionnaire items.   The nature of this association can be better studied after the observed frequencies are converted to proportions by columns (Table 3.20).

TABLE 3.20　　　*Conversion to proportions by columns*

		Agree	*Neutral*	*Disagree*	*Total*
			Item A		
	Agree	.625	.333	.357	.400
Item *B*	Neutral	.250	.333	.143	.250
	Disagree	.125	.333	.500	.350
	Total	1.000	1.000	1.000	1.000

3.3　Chi-square test for experimental homogeneity

When data are available on one categorical variable for two or more random samples, the chi-square statistic may be used for a test concerning the equivalence, or homogeneity, of the populations from which the samples were drawn. The null hypothesis and computation of a test statistic are the same for a chi-square test of independence. The interpretation of the result differs only in that a statement comparing two or more populations can be made. Consider an example of an experimental comparison of three types of guidance services, where the outcome variable of interest is retention in school. Specifically, a population of potential public school dropouts is identified among the ninth graders in a large school district. Three random samples of 60 students each are chosen from this population. The first sample is provided with special in-school guidance services during their high school years; the second sample is provided with at-home counseling in addition to the same special in-school guidance given to the first group; the third sample receives no special guidance service of any type. At the end of a 3-year period, the numbers of those who dropped out of school and were retained in school are tallied for the three samples. It is of interest to decide whether the three guidance procedures produced significant differences in the dropout rates among the groups of potential dropouts. The data may be summarized in the form of a two-way classification table with the columns representing the three experimental treatment groups and the rows representing the dichotomized variable "dropped out or retained in school" (Table 3.21). By use of conventional contingency table notation, the null hypothesis is written

$$H_0: \begin{cases} \pi_{11} = \pi_{12} = \pi_{13} \\ \pi_{21} = \pi_{22} = \pi_{23} \end{cases}$$

TABLE 3.21   Observed frequencies for guidance data

| | Experimental treatment | | | |
	None	In-school guidance	In-school + at-home guidance	Total
Dropped out	40	31	25	96
Retained	20	29	35	84
Total	60	60	60	180

In words, this null hypothesis states that the proportions of dropouts would be the same for populations receiving the three types of guidance services. The estimated proportion of those dropping out is $96/180 = .533$. Also, the proportion retained in school is estimated as $.467$. Expected frequencies under H_0 are shown in Table 3.22. The computed value of the χ^2 statistic is

$$\chi^2 = \frac{40^2}{32.00} + \cdots + \frac{35^2}{28.00} - 180 = 187.64 - 180 = 7.64$$

This χ^2 value has $(k - 1)(c - 1) = 2$ degrees of freedom. We see from Table F.4 that the probability of an outcome as extreme as or more extreme than the observed is between .025 and .01. Thus, at the conventional .05 level of significance, we would reject the null hypothesis and decide that the dropout rates

TABLE 3.22   Observed and expected frequencies for guidance data

	None	In-school guidance	In-school + at-home guidance	Total
Dropped out	40/32.00	31/32.00	25/32.00	96
Retained	20/28.00	29/28.00	35/28.00	84
Total	60	60	60	180

were significantly different for the three treatment groups. That is, the three experimental treatments did not produce results which were homogeneous with respect to dropout rate. Inspection of the observed frequencies in Table 3.21 indicates that the group with no guidance showed a dropout rate of 67 percent, whereas the two special counseling groups showed 52 and 42 percent, respectively.   Since the third group had special at-home as well as in-school guidance, it would also be of interest to compare this group with the second group that had special in-school guidance only.   This can be done by setting up a 2 × 2 table for just these two groups.   Table 3.23 shows the observed frequencies for the

TABLE 3.23   *2 × 2 summary table for experimental treatments*

	In-school guidance	In-school + at-home guidance	Total
Dropped out	31/28.00	25/28.00	56
Retained	29/32.00	35/32.00	64
Total	60	60	120

two groups as well as the expected frequencies under the null hypothesis H_0: $\pi_{11} = \pi_{12}$.   The corrected chi-square value is found to be $\chi^2 = .84$.   With 1 degree of freedom, this value has a probability greater than .05.   Thus, we conclude that the two guidance programs do not produce significantly different dropout rates.   To summarize the results of the *two* chi-square tests, we could state that "At-home plus in-school guidance did not produce a significantly lower dropout rate than did in-school guidance alone.   However, when compared with a no-guidance program, significant differences favoring the use of some form of guidance were found."

3.4   *Short-cut computational routines*

For certain types of two-way classification tables, simplications may be introduced into the computational formulas for the chi-square statistic.   Two such shortcut procedures are of special interest to the researcher, since they apply to types of table commonly encountered.

1)   Special 2 × 2 chi-square formula.   *Observed* frequencies in a 2 × 2 contingency table can be used to compute χ^2 directly *without* first finding expected frequencies.   The modified formula, including the correction term, is

$$\chi^2 = \frac{f_{..}(|f_{11}f_{22} - f_{12}f_{21}| - f_{..}/2)^2}{f_{1.}f_{2.}f_{.1}f_{.2}}$$

where the symbols represent observed frequencies from the standard 2×2 contingency table.

	Column variable		
	1	2	Total
Row 1	f_{11}	f_{12}	$f_{1.}$
Variable 2	f_{21}	f_{22}	$f_{2.}$
Total	$f_{.1}$	$f_{.2}$	$f_{..}$

To illustrate the use of this formula, consider the data from Table 3.23. Substituting the observed frequencies in the formula, we obtain

$$\chi^2 = 120 \frac{(|31 \cdot 35 - 25 \cdot 29| - 120/2)^2}{56 \cdot 64 \cdot 60 \cdot 60} = \frac{120 \cdot 300 \cdot 300}{56 \cdot 64 \cdot 60 \cdot 60} = .84$$

Note that this result is in agreement with that obtained when computation was based on both observed and expected frequencies.

2)   Special $2 \times c$ chi-square formula.   When one of the two marginal distributions is dichotomized, we can set up the table with two rows and c columns.   The simplified formula utilizes the following notation for *observed* frequencies:

	Column				
	1	2	. . .	c	Total
Row 1	f_{11}	f_{12}	. . .	f_{1c}	$f_{1.}$
Row 2	f_{21}	f_{22}	. . .	f_{2c}	$f_{2.}$
Total	$f_{.1}$	$f_{.2}$		$f_{.c}$	$f_{..}$

Then,

$$\chi^2 = \frac{f_{..}^2 \left(\sum_{j=1}^{c} f_{1j}^2/f_{.j} \right) - f_{..}f_{1.}^2}{f_{1.}f_{2.}}$$

As was true of the special formula for 2 × 2 tables, this formula avoids the computation of expected frequencies.   To illustrate the 2 × c formula, the observed frequencies from Table 3.21 can be substituted to obtain

$$\chi^2 = \frac{180^2(40^2/60 + 31^2/60 + 25^2/60) - 180(96)^2}{96(84)} = \frac{61,560}{8,064}$$

$$= 7.63$$

This result agrees, within rounding error, with the result previously obtained.

The student is urged to utilize the two shortcut formulas whenever they are appropriate to save himself computational labor.

3.5   *Limitations on the use of chi-square tests*

The chi-square tests for contingency table data are generally quite easy to set up and are widely used in behavioral science research.   However, tendencies toward certain types of misapplications of chi-square tests are also prevalent, and the student should be aware of definite limitations that exist for this statistical test. †

1)   All of the formulas presented for chi-square statistics are in terms of frequencies.   Do not attempt to compute on the basis of proportions or other derived measures.

2)   Frequencies of nonoccurrence should not be omitted for binomial or multinomial events.   For example, if 4 drugs were tried out on 4 separate groups of 100 patients each, the number of cures per drug might be shown in a one-way table (Table 3.24).   However, a chi-square test should not be applied to these data until the alternative outcome (i.e., "not cured") is represented in the table (Table 3.25).

TABLE 3.24   *Data for four drugs*

	Drug				
	1	*2*	*3*	*4*	*Total*
Number cured	30	60	40	80	210

† For the student interested in some of the history and controversy which have surrounded chi-square tests, see Lewis and Burke [1949 and 1950], Pastore [1950], Edwards [1950], and Peters [1950].

TABLE 3.25   *Contingency table for drug data*

	Drug				
	1	*2*	*3*	*4*	*Total*
Number cured	30	60	40	80	210
Number not cured	70	40	60	20	190
Total	100	100	100	100	400

3)   The formulas presented in this chapter are not appropriate for cases in which repeated measurements on the same or matched groups are represented in one table.† Techniques suitable for such data are available and the student should consult Siegel [1956, pp. 63–67 and 161–166] and Cochran [1950].

4)   The accuracy of the probabilities from corrected χ^2 based on 2×2 tables is quite poor when small samples are involved. As a general rule, apply the chi-square test to 2×2 tables only for samples larger than 40. For sample sizes less than 40, use the exact test based on the hypergeometric distribution [Fisher, 1946; Siegel, 1956]. However, for the special case in which one or both of the marginal distributions are *exactly* evenly divided (that is, $f_{11} + f_{12} = f_{..}/2$ and/or $f_{11} + f_{21} = f_{..}/2$), the corrected χ^2 value is accurate for any sample size (Dayton, 1964).

5)   For tables with more than two rows or two columns, the rules offered by Cochran [1954] provide sensible guidelines. Cochran recommends that the chi-square test be applied to $k \times c$ tables as long as all *expected* frequencies are greater than 1 and not more than 20 percent of the *cells* show *expected* frequencies less than 5. For a 3×4 contingency table, this rule would allow for no more than two expected frequencies less than 5

† When data from questionnaires and similar devices are analyzed, the student should be careful that he does not set up the tables incorrectly. For example, it may seem reasonable to set up a table as follows:

	Agree	Neutral	Disagree	Total
Item *A*	40	90	70	200
Item *B*	80	50	70	200

However, a chi-square contingency test should not be performed on the basis of this table, since it is not really a *contingency* table. Note that the sample included only 200 students, yet this table shows a total of 400. This is so because each student is classified *twice* in the table.

(but greater than 1). In a 5 × 6 contingency table as many as six cells could show expected frequencies less than 5 (but greater than 1).

Problems

3.1. In a group of 150 students, there are 50 who are male and 30 who got a grade of A on a test. Assuming grades to be independent of sex, how many females would you expect to get grades other than A?

3.2. In a freshman mathematics class of 100 students, there are 60 males and 40 females. The distribution of grades at the end of the semester is shown below. Compute the phi coefficient to determine if there is any relationship between sex and grades for the population of students.

		Sex	
		Male	Female
Grade	A	20	5
	Other than A	40	35

3.3. Use the data in Prob. 3.2 and assume that it represents a random sample of 100 students. Test the hypothesis $H_0: \pi_{A_M} = \pi_{A_F}$. Use the .05 level of significance.

3.4. In a study of the reading ability of 50 bilingual and 50 unilingual college students, the observed scores on a reading comprehension test were recorded as shown below. Analyze the data to determine if reading ability is independent of language background. Use the .01 level of significance.

		Language background	
		Bilingual	Unilingual
Reading comprehension	High	20	17
	Moderate	18	20
	Low	12	13

3.5. How are the degrees of freedom most easily obtained for chi-square tests of independence?

3.6. Special individualized reading instruction was given to 145 children for 5 hours each week. When classified by mental age (above and below a specified level) and whether or not reading improvement was noted, they were distributed as below. Test the hypothesis of independence using the .05 level of significance.

		Mental ability	
		Low	High
Improvement	Yes	40	40
	No	40	25

3.7. For a sample of 30 graduate students who took a foreign language reading examination for the first time, their success is tabled in relation to whether they prepared for the examination through formal instruction or by self-study. Using the .05 level of significance, test the hypothesis of independence.

		Outcome of examination	
		Passed	Failed
Preparation	Formal instruction	7	13
	Self-study	9	1

3.8. A teacher devised three different methods for presenting an abstract concept to students. After randomly assigning 50 students among 3 groups, instruction was given utilizing the methods and the students were then tested to determine whether or not the concept had been mastered. The results are tabled below. Using the .05 level of significance, decide whether or not the instructional methods were equally effective.

		Instructional method		
		A	B	C
Outcome	Learned concept	8	7	5
	Did not learn concept	12	3	15

Four

descriptive methods for univariate measured data—one group

4.1 Introduction

Data that are based on measured variables offer the widest and most complete possibilities for analysis and interpretation. For this reason, it is desirable, whenever possible, to select measuring instruments and procedures that yield equal units. In the behavioral sciences, the equal-unit criterion is sometimes difficult to meet with available measuring devices. There are techniques within the field of psychometric methods for investigating whether or not a given measurement procedure does in fact result in equal units along the measured variable; these techniques are, however, too advanced for presentation in an introductory textbook. For the beginner in educational research, the following advice is appropriate: data should be treated as interval *only if* the existence of equal units can be defended on logical or empirical grounds; when this cannot be done, it may still be possible to justify the use of techniques for ordinal data if it can be shown that the data at least rank students accurately. To illustrate the types of questions you might raise concerning the nature of different kinds of data, some examples will be discussed.

Probably the most common kind of data encountered in educational research is that derived from some form of achievement measure. Assume that a unit of study at the elementary school level is concerned with the spelling of 25 three-syllable words. At the end of the unit of study, the teacher examines the students by reciting the words one at a time, in some arbitrary order, to the class. The student's task is to write each word as it is recited to him. A score for a student is obtained by counting the number of words correctly spelled by that student. Thus, the maximum score is 25 and the minimum is 0. Can scores of this type justifiably be treated as representing equal units of measurement?

An answer to this question depends upon the definition of the variable involved in the measurement. If the variable is defined as simply the number of words correctly spelled under the prevailing conditions of testing, then the scores do represent equal units, and an absolute zero point, in fact, exists. Thus, the data are on a ratio scale with respect to this variable. On the other hand, if the variable is defined more generally as the amount learned by the student concerning the spelling of the 25 words, it can be argued that equal units do not prevail. To begin with, it is unlikely that the 25 words are equally difficult to spell. Thus, one score unit may represent a more difficult learning task than some other score units. Also, since some students may have been able to spell certain of the words before the beginning of the instructional unit, their amount of new learning could be less than other students achieving equal, or even lower, scores on the examination. We most likely would have to conclude that the data do not represent equal units for the variable "amount learned." We might, however, be satisfied that the students are correctly ranked along this variable.

Consider another example of educational measurement. As part of a project on teacher attitudes toward students, an instrument is developed and administered to a group of teachers. For each of the items the teacher indicates his opinion by marking a 5-point preference scale. An exemplary item is:

Children who present extreme disciplinary problems should not be punished, since their behavior stems from deep-rooted psychological problems.

A	B	C	D	E
Strongly agree	Agree	Neutral	Disagree	Strongly disagree

A numerical score for each item could be obtained by letting A be 5, B be 4, C be 3, D be 2, and E be 1. It is reasonably apparent that it would be difficult to establish that these data represent equal units of measurement with respect to an attitudinal variable. It should be noted that data similar to those described in the last example are sometimes analyzed by techniques that are appropriate only to interval and ratio data. These procedures may be justifiable in specific cases; but the student is urged to maintain a skeptical attitude concerning these applications, since they hinge upon subtle arguments that require the advice of a professional statistician.

The variety of available analytical tools for data on an interval or a ratio scale of measurement is attributable to the fact that such data can be algebraically manipulated in a meaningful way. That is, the usual arithmetic operations—addition, multiplication, etc.—can be applied to interval or ratio data and the results of such operations can be interpreted directly. On the other hand, nominal and ordinal data can be treated mathematically only in certain limited ways (e.g., frequencies of occurrence can be added, and so forth).

The next four chapters are devoted to procedures applicable to data measured on interval and ratio scales. In this chapter, the case of description for one group of subjects and one measured variable will be considered. Chapter 5 extends the coverage of the one-group case to inferential techniques. Descriptive and inferential procedures for two or more groups are then considered in Chap. 6. Finally, Chap. 7 is devoted to bivariate analysis for one group of subjects.

4.2 Tabular and graphical procedures

Frequency distributions

As was the case with categorical data, the first techniques we shall consider involve summarizing data in the form of tables or graphs. In most cases, whole numbers (or integers), rather than decimal numbers, will be used in examples. This is done because most measurements arising in educational research are, in fact, whole numbers. Also, the typical example will concern the analysis of scores derived from testing instruments of various sorts, since this type of data is so commonly encountered in educational research. At this point, it becomes important to recall and review some of the concepts related to measurement that were treated in Chap. 1 and, also, to introduce the notion of a measurement unit. As noted earlier, variables may be discrete or continuous. The numbers resulting from measurement are, in either case, discrete quantities in the sense that there is always some practical limitation to the accuracy with which measurement is carried out. Thus, the variable "intelligence" is usually thought of as continuous, and all shades of difference among people along this variable are possible. In practice, however, scores from intelligence tests are expressed as whole numbers and no finer measurement is attempted. That is, an intelligence score might be 105 or 106, but 105.5 would not be reported. Many variables involved in educational research are reported in terms of whole numbers, but notable exceptions do exist. Grade-equivalent scores, which are derived from some standardized achievement tests, are reported as decimal numbers, with the whole number portion standing for grade and the decimal portion for month of the school year. Thus, a grade equivalent of 3.8 means the eighth month of the third grade. Also, teachers sometimes allow fractional points on classroom tests and scores such as 76½ are possible.

The term *measurement unit* refers to the smallest difference between scores that is measured and reported. In the case of intelligence-test scores, the measurement unit is 1 score unit, since the scores are reported as consecutive integral values. A grade-equivalent score, on the other hand, has a measurement unit of .1 school year (or 1 month, since a school year is 10 months). To illustrate the concept of measurement unit in more detail, let us consider examples from physical measurement. The lengths of objects are often measured in inches or fractions of inches. The measurement unit can usually be inferred from the measures themselves. Thus, the measurement 8.14 inches implies a measurement unit of .01 inch, and the measurement 24.00214 inches implies a measurement unit of .00001 inch. For most physical variables, the accuracy of measurement can be increased to very high levels; hence, the measurement unit may be an extremely small quantity.

For continuous variables, the meaning of an individual score needs to be examined in more detail. If the length of an object is reported as 21.2 inches, we know that the measurement unit is .1 inch, since differences as small as .1 inch were measured. Also, the length 21.2 inches implies that the object was nearer 21.2 inches than it was to 21.1 inches or 21.3 inches. That is, the person making the measurement has concluded that the object's length is 21.2 inches to the *nearest .1 inch*. If more accurate measurement had been carried out, the length would perhaps have been 21.18 inches or 21.23 inches. These latter figures, of course, could also have been improved upon by increased sensitivity of the measurement procedure. For the measurement 21.2 inches, any length just over 21.15 inches or just less than 21.25 inches would be reported as 21.2 inches if the measurement unit were .1 inch. Points such as these are called the *true limits* for a score. As another example, the weight 102 pounds is based on a measurement unit of 1 pound; and its true limits are 101.5 pounds and 102.5 pounds, since any measurement between these points would be reported as 102 pounds. As a general method, the true limits can be found by first adding to and then subtracting from the measurement, the quantity $\frac{1}{2}$ (measurement unit).

The basic purpose of tabular and graphical procedures is to take quantities of numerical data and organize them into forms that facilitate communication and interpretation of these data. From the point of view of communicating data to other interested persons, it is obvious that as the number of scores increases, a simple listing of these scores will prove of little value. Also, the process of interpreting the meaning of groups of scores becomes more burdensome as larger numbers of scores are involved. Imagine attempting to draw conclusions from a list of 1,000 or more intelligence-test scores for students in one junior high school.

The basic device for organizing large groups of data into manageable form involves grouping scores into intervals rather than reporting individual scores. That is, the number, or *frequency*, of scores in specified intervals along the score dimension is determined. A table is then formed with these frequencies recorded for each interval of scores. The manner of selecting intervals and procedures for determining the frequencies are illustrated for the score data given in Table 4.1. These data represent raw scores (maximum score possible was 40) on a quiz given to 55 students enrolled in an introductory educational statistics course. The scores are listed in the order in which they appear in the instructor's grade book; that is, they are arranged according to the alphabetical listing of students in the statistics class. Note that even a relatively small number of scores, such as 55, presents a rather bewildering accumulation of data when presented in the form used in Table 4.1. These data may be reduced to a convenient tabular form by choosing score intervals and reporting frequencies per interval.

TABLE 4.1   Quiz scores for 55 statistics students

35	16	19	33	40
29	29	29	16	35
23	25	27	28	38
36	32	15	27	36
38	39	40	32	31
36	33	39	21	24
40	36	30	32	25
36	33	40	36	30
39	32	34	28	40
33	35	40	26	37
38	35	31	38	39

When selecting the score intervals, several principles should be kept in mind: (1) keep the number of *measurement units* in the intervals a whole number and constant; that is, combine 3 measurement units or 4 measurement units but not $3\frac{1}{2}$ measurement units in each interval; (2) make all intervals adjacent to one another, even though gaps may exist in the distribution of actual scores; (3) in general, use between 8 and 20 intervals—for large numbers of scores use nearer 20 intervals, for small numbers, use nearer 8 intervals. The object here is to use enough intervals so that the data are accurately represented but not so many that the table becomes cumbersomely large. That is, a balance must be maintained between the amount of information retained and the conciseness and compactness of the resulting table. From practical experience, 8 to 20 intervals seem to be reasonable for most groups of scores.

For the example of the 55 quiz scores, a relatively small number of intervals would suffice, say 8 or 9. To set up these intervals, scan the distribution to find the smallest and largest scores. The smallest score, denoted X_{min}, is 15; and the largest score, denoted X_{max}, is 40. Thus, the scores vary from 15 to 40. The number of units from the smallest to the largest score is defined as the *range* of the scores and is symbolized by w. The range can be computed from the formula $w = X_{max} - X_{min} + 1$, when scores are whole numbers.† For the example, $w = 40 = 15 + 1 = 26$. To divide the range into, say, 9 intervals, each interval should contain $2\frac{6}{9}$ or $2\frac{8}{9}$ score units. Since only a whole number of score units is generally included in an interval, the best choice is to include 3 score units per interval. The number of score units included in an interval is called the *interval width* and is denoted i; thus, $i = 3$ for this example. The construction of the actual score intervals can now proceed. The first interval may be chosen as either 13–15, 14–16, or 15–17, since any one of these includes the smallest score, 15. While it makes little difference which first interval is used, it is conventional to let X_{min} fall at, or near, the middle of the first interval.‡ Since $X_{min} = 15$ is the middle of the interval 14–16, we would choose this as the first interval. Once the first interval is found, the remaining intervals can be constructed quite mechanically by repeatedly adding i to the limits for the

† Some textbook authors prefer to define the range simply as $w' = X_{max} - X_{min}$. However, this formula fails to account for both endpoints of the score distribution. Consider the example of just three scores, 4, 5, and 6. Our formula yields $w = 6 - 4 + 1 = 3$. The alternative formula gives $w' = 6 - 4 = 2$. The three scores, 4, 5, 6, however, cover 3, not 2, score units; and this illustrates the rationale for our preference. For scores in a form other than whole numbers, the formula must be modified to become $w = X_{max} - X_{min} + 1$ measurement unit. Thus, if scores are recorded to the nearest tenth (e.g., 8.3, 7.9, 5.2, etc.), the formula is $w = X_{max} - X_{min} + .1$.

‡ One should, however, always exercise common sense in constructing the intervals. If, for example, the data are achievement-test scores and $X_{min} = 0$, the first interval might be chosen as 0 — 4; placing X_{min} at the center of the interval would result in a negative lower limit, but negative scores would be impossible on the achievement test.

previous interval. Thus, the second interval is 17–19, since $14 + 3 = 17$ and $16 + 3 = 19$. Table 4.2 displays the total set of intervals. Note that no interval beyond 38–40 is included, since this interval includes X_{max}, or 40.†

TABLE 4.2   Score intervals for 55 quiz scores

Scores	
38–40	23–25
35–37	20–22
32–34	17–19
29–31	14–16
26–28	

Now that the score intervals have been formed, the number of scores per interval can easily be found by tallying the scores from Table 4.2 into the score intervals. A table displaying score intervals and the frequency per interval is called a *frequency distribution*. Table 4.3 shows the frequency distribution (and also the tally) for the 55 quiz scores.

TABLE 4.3   Frequency distribution for 55 quiz scores

Scores	Tally	Frequency
38–40	⦀⦀ ⦀⦀ ////	14
35–37	⦀⦀ ⦀⦀ /	11
32–34	⦀⦀ ////	9
29–31	⦀⦀ //	7
26–28	⦀⦀	5
23–25	////	4
20–22	/	1
17–19	/	1
14–16	///	3
Total		55

The procedure given for constructing a frequency distribution is applicable to the majority of situations involving educational research data. However, special types of data require certain modifications.

† Conventionally, the intervals are written so that large-score values are at the top of the table and small-score values are at the bottom of the table.

(1) If scores are not whole numbers or if the measurement unit is larger than 1, care must be exercised in determining the limits for the score intervals. Consider first the situation in which measurements are recorded as decimal or fractional numbers (see data in Table 4.4). While the steps for forming a frequency distribution are no different from those for whole-number data, one must be careful to take into consideration the existence of the decimal point. The maximum and minimum scores are $X_{max} = 20.0$ and $X_{min} = 5.2$. The range is $w = X_{max} - X_{min} + .1$, or $w = 20.0 - 5.2 + .1 = 14.9$. Note that the formula for w has been modified to account for the unit of measurement. If about 10 intervals were desired, the interval width would be $14.9/10 = 1.49$, or $i = 1.5$. A reasonable choice for a first interval is 4.5–5.9, since this contains X_{min}, and the lower limit of the interval provides a convenient starting point for constructing additional intervals. The first few intervals would be 4.5–5.9, 6.0–74, 7.5–8.9, 9.0–10.4, 10.5–11.9, etc. The student should confirm that we actually end up with 11 intervals in order to cover all scores from 5.2 through 20.0. The fact that we have 11 rather than the intended 10 intervals should not be disturbing, since this falls well within the recommended number of intervals (that is, 8 to 20).

TABLE 4.4   Coded reading scores for
30 students

14.8	7.4	17.9	16.3	9.9
13.2	8.3	11.4	6.4	11.3
11.9	11.9	12.8	20.0	15.0
14.6	14.2	16.1	15.3	8.6
15.1	15.1	14.2	12.8	19.4
14.6	12.2	18.7	5.2	19.2

The case in which the unit of measurement is larger than 1 does not occur frequently in educational research, but it is worth illustrating. The data in Table 4.5 are estimates by a classroom teacher of the family incomes for the 22 students in her class. It is unrealistic to expect the teacher to estimate to the nearest dollar or even to the nearest $10; therefore, she estimated the income information in $100 units. The unit of measurement is thus $100; and the range is $w = X_{max} - X_{min} + 100$, or $w = 12,500 - 3,2000 + 100 = 9,400$. If about 8 intervals were desired, an interval width of $1,200 is reasonable, since $9,400/8 = 1,175$; rounding this to the nearest $100 yields $1,200. If the first interval is taken as 2,500–3,600, then the next few score intervals are 3,700–4,800, 4,900–6,000, 6,100–7,200, 7,300–8,400, etc.

TABLE 4.5   Estimated family dollar incomes for 22 students

6,800	4,600	12,500
4,500	9,400	6,700
8,200	5,300	3,500
6,500	6,100	4,700
10,000	4,700	6,100
5,800	5,400	5,600
5,200	5,800	
3,200	8,600	

(2) In some distributions of scores, there may be a few scores that are far removed from the remainder of the scores. For example, the distribution of intelligence-test scores for college professors at a certain institution may show most scores at 130, 140, or above; however, there may be one or two quite low scores, at 90 or 95, for example. For constructing score intervals, if the deviant scores are used in determining the interval width, this may result in several empty intervals toward the low end of the distribution. It is often more convenient in practice to treat these deviant scores in a special way and thereby to construct a more compact frequency distribution. This special device can be most easily understood by reference to Table 4.6. The highest interval is labeled "3.2 or

TABLE 4.6   Reading-achievement–grade-equivalent scores for 85 homo-geneously grouped second graders

Scores	Frequency
3.2 or greater	2
3.0–3.1	3
2.8–2.9	7
2.6–2.7	8
2.4–2.5	12
2.2–2.3	14
2.0–2.1	13
1.8–1.9	10
1.6–1.7	8
1.4–1.5	7
1.2–1.3	1
Total	85

greater," indicating that two deviant scores exist; in fact, these are 3.8 and 4.6. In order to accommodate these two scores in the table, it would have been necessary to add 8 additional score intervals, 6 of which would contain frequencies of 0. The advantage of the interval "3.2 or greater" is that one has a more compact table without sacrificing information concerning the majority of scores, which are between 1.2 and 3.1. An interval that contains no upper-score limit (or lower-score limit) is referred to as an *open interval*. Thus, "3.2 or greater" is an open interval at the high-score end of the table. An open interval at the low-score end of a table might read "$\frac{1}{4}$ or less" or "99 and smaller."

A word of caution concerning the use of open intervals is appropriate. Unless the data dictate the use of an open interval, they should be avoided. The disadvantage of open intervals is the extreme loss of information concerning scores in these intervals. For the open interval in Table 4.6, you were told that the two scores were actually 3.8 and 4.6. However, in most practical situations, the reader of such a table will have no way of knowing these actual values.

An alternative to the use of an open interval is to define an interval that is wider than the remaining ones in the tables. For the example, the final interval could have been 3.2–4.6 with a frequency of 2. Although this recommendation violates the rule that interval widths should be constant, it is an acceptable alternative to the use of an open interval, since less information is lost concerning the extreme scores.

Frequency polygon and histogram

The information contained in a frequency distribution may be presented in graphic form in several different ways. The simplest technique is to construct a line graph showing the frequency for each score interval; such a graph is known as a *frequency polygon*. A second type of graph—somewhat more difficult to construct—is the *histogram*, in which frequency per interval is represented by rectangles or bars.

A frequency polygon may be constructed from any frequency distribution that does not contain open intervals. The graph is built on ordinary Cartesian, or rectangular, coordinates with the vertical axis marked off in terms of frequency and the horizontal axis marked off in terms of scores. The frequency per interval is then represented by a point above the midpoint of the score interval at a distance equal to the frequency for that interval. The resulting points are connected by line segments to produce the completed graph. Since the midpoints of the score intervals are used for plotting purposes, it is common to compute these midpoints directly from the frequency-distribution table and plot them, rather than the score limits, along the horizontal axis of the graph. These midpoints, known as the *class indexes* (or class marks), may be computed systematically by averaging the upper- and lower-score limits for the first score interval. For the data in Table 4.3, the first interval is 14–16 and the average of these is $(14 + 16)/2$, or 15. The class index for this interval is thus 15. The remaining class indexes may be found by repeatedly adding i, the interval width, to the value found for the first interval. Thus, the second class index (for the interval 17–19) is $15 + 3$, or 18; the third is $18 + 3$, or 21; and so forth through the last interval, 38–40, which has a class index of 39. Table 4.7 displays the original score intervals, the class indexes, as well as the frequency per interval for this example. Now that the class indexes have been found, the data in Table 4.7 will be used to illustrate the construction of the actual frequency

TABLE 4.7   Frequency distribution
for 55 quiz scores

Score	Class index	Frequency
38–40	39	14
35–37	36	11
32–34	33	9
29–31	30	7
26–28	27	5
23–25	24	4
20–22	21	1
17–19	18	1
14–16	15	3
Total		55

polygon. The coordinate system shown in Fig. 4.1 can be constructed with ordinary graph paper and a ruler. Starting at 0, frequency is plotted along the vertical axis in convenient units. Note that the maximum frequency shown is 14, since no larger frequency appears in Table 5.7. The class indexes have been plotted along the horizontal axis. Note, however, that one additional class index has been added at each end of the axis (i.e., the class index 12 at the left and the class index 42 at the right). The reason for this will become apparent when the line segments are drawn. Also, the horizontal axis below 12 is indicated by a jagged line.

The next step in constructing the frequency polygon involves plotting the frequencies as points above the corresponding class indexes and connecting these points by line segments. (See Fig. 4.2.) Note that the right and left extremes of the graph have been closed off by the segments connected with the score axis. This was the reason for including the extra class indexes 12 and 42. The name

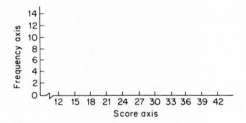

figure 4.1   *Axes for constructing a frequency polygon.*

"frequency polygon" is derived from the fact that the completed graph, considering the line segments *and* the score axis, is in the shape of a many-sided figure, or *polygon*.

A frequency polygon, while containing no more information than a frequency-distribution table, is often useful for visually studying the shape of a distribution of scores. The preponderance of high scores is immediately apparent from Fig. 4.2. A conclusion from a study of the frequency polygon (or the frequency distribution, for that matter) would be that this quiz was relatively "easy" for the students who took it.

An alternative graphical procedure, equivalent in information to the frequency polygon, is the histogram. A histogram is essentially a bar graph with the bars, or rectangles, constructed over the score intervals with heights equal to the frequency for the corresponding intervals. Since histograms are generally used for score dimensions that represent continuous variables, it is conventional to construct the rectangles contiguous to one another (see Fig. 4.3 for a histogram based on the 55 quiz scores). In order to eliminate gaps, the rectangles are constructed not on the score limits for the intervals but halfway between adjacent upper- and lower-score limits. For the case of continuous variables, the logic behind this process is straightforward. An interval of scores has true limits in the same sense that an individual score does. The interval 14–16, for example, contains any score between 13.5 and 16.5. Thus, 13.5 and 16.5 are the true

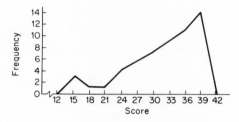

figure 4.2   *Frequency polygon for 55 quiz scores.*

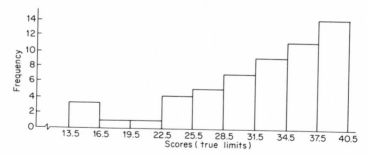

figure 4.3   Histogram for 55 quiz scores.

limits for the score interval 14–16.   In order to fill up the gap between adjacent
score intervals, the rectangles of the histogram are constructed on the true limits
for the score interval.   The true limits for a score interval may be computed
mechanically by the following procedure: compute $i/2$; subtract $i/2$ from the
class index for an interval (this yields the lower true limit for the interval); and
add $i/2$ to the class index for an interval (this yields the upper true limit for the
interval).   The score intervals and true limits are shown for the frequency distribution of 55 quiz scores in Table 4.8.   The histogram in Fig. 4.3 was constructed by plotting the true limits from Table 4.8 along the horizontal axis and
then constructing rectangles on these true limits at a height corresponding to the
frequency for each interval.

As was pointed out earlier, the notion of true limits is based on the assumption that the variable under consideration is continuous.   For variables that
are discrete (as is the variable "number of correct responses on a quiz"), true

TABLE 4.8   *Frequency distribution for 55 quiz scores*

Score intervals	True limits	Frequency
38–40	37.5–40.5	14
35–37	34.5–37.5	11
32–34	31.5–34.5	9
29–31	28.5–31.5	7
26–28	25.5–28.5	5
23–25	22.5–25.5	4
20–22	19.5–22.5	1
17–19	16.5–19.5	1
14–16	13.5–16.5	3
Total		55

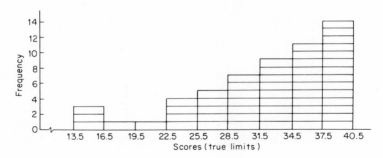

figure 4.4   Histogram showing area units.

limits for scores are meaningless, since only specific, discrete values of the variable are possible.   Nevertheless, it is *conventional* in educational reporting to use true limits for data based on discrete variables.   Thus, the histogram in Fig. 4.3 must be viewed as a conventional manner of presenting data that violates the requirements for a continuous variable.   Since this same problem does not arise when constructing frequency polygons,† and since frequency polygons are easier to construct, this graphic form is recommended to the student.

An interesting interpretation can be made of the area included in a histogram or frequency polygon.   Since the interpretation is more apparent for the case of the histogram, we shall start with that type of graph.   First, define a "unit of area" as the rectangular area with a base of i score units and a height of 1 frequency unit.   For the histogram based on the 55 quiz scores (Fig. 4.3) this unit has a base of 3 score units and a height of 1 frequency unit.   Any one of the rectangles, or bars, comprising the histogram can be conceived of as 1 or more of these area units piled on top of one another.   Figure 4.4 shows this for the 55 quiz scores.   The number of area units required for any given rectangle is the same as the frequency of scores in that interval.   Thus, the frequency with which scores occur in a score interval can be interpreted as a corresponding number of units of area.   For example, the frequency of scores in the score interval 26–28 from Table 4.7 is 5.   On the histogram for these data, the rectangle representing the frequency for this interval has an area of 5 units.   In a broader sense, it can be said that the area of the entire histogram is 55 units, since the total frequency of scores upon which the graph is based is 55.   Similarly, we might speak of the area in the histogram above, below, or between various scores or true limits for scores.   The area above the true limit of 34.5 is 25, since the total frequency of scores above 34.5 is 25.   Also, the area below 22.5 is 5; the area between 22.5 and 34.5 is 25.   That a similar relationship exists between

† The fact that a class index for an interval representing scores along a discrete variable may be an impossible value (e.g., the interval 14–17 has a class index of 15.5) should be recognized; here again, we *conventionally* represent the interval by the average of its score limits.

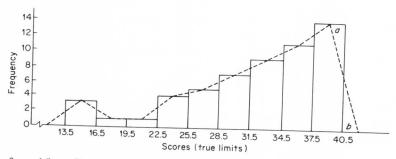

figure 4.5   *Frequency polygon superimposed on histogram for 55 quiz scores.*

area and frequency in a frequency polygon can be seen from Fig. 4.5.   Here, the frequency polygon for 55 quiz scores has been superimposed on the histogram.   Certain areas under the frequency polygon are not included under the histogram (e.g., the area labeled *b*); also certain areas under the histogram are omitted under the frequency polygon (e.g., the area labeled *a*).   It is apparent, however, that these omitted areas compensate for one another, so that the total area under the two graphs is the same (i.e., the area labeled *a* is equivalent to the area labeled *b* and they compensate for one another).   Thus, the total area under the frequency polygon is also 55 units.   An interesting extension of the notion of interpreting area as frequency is possible if the frequencies from a frequency-distribution table are first converted to proportions and then these proportions are plotted in the form of a histogram or frequency polygon.   The resulting figure will have the same horizontal axis (i.e., scores) as an ordinary histogram or frequency polygon; the vertical axis, however, would be marked off in units representing proportions.   In such a figure, the proportion of cases above or below a given score, or the proportion of cases between two scores, is represented by the corresponding area.   The concept of frequency as area in a graph will prove useful to us in later chapters where it is elaborated on in more detail.

Cumulative proportion distribution

It is often of interest to make statements concerning the relative position of scores in a distribution of scores.   This is the case when standardized examinations are used for achievement testing or for selection and classification purposes with school children.   The raw scores of students on such tests are of little interest; what is desired is some comparison of individual students or groups of students with regional or national groups, known as norm groups.   Thus, a second-grade teacher may wish to know how well her pupils read compared with national (or state or regional) norm groups of comparable students.   In order to uncover this information, a standardized reading test may be administered to the classroom of students and their individual scores used to determine where each student stands relative to the norm group. In this section, tabular and graphic procedures are presented for ascertaining the relative positions of scores.

Let us begin with some terminology. One very useful measure of the relative position of a score in a distribution of scores is the percentage of scores equal to or less than the given score. This percentage figure is known as the *percentile rank* of the score. A notion closely related to percentile rank is that of *percentile*. A percentile is a *score* at or below which a stated percentage of the cases fall. The term percentile is used with some specific numerical prefix. For example, the 75th percentile is the score at or below which 75 percent of the scores in a distribution fall. It is quite easy for students to confuse the terms percentile and percentile rank. It is necessary to realize that a percentile is always a point on the score axis, whereas a percentile rank is always a number representing a percentage. Thus, if in a certain distribution, 75 percent of the scores fall at or below a score of 24, then 24 is the 75th percentile of the distribution and the percentile rank for the score of 24 is 75.

Percentiles are denoted by the symbol P, with a subscript corresponding to the percentage of scores below that percentile. Thus, the 75th percentile would be written P_{75}, the 30th percentile would be written P_{30}, and so forth.

Percentiles and percentile ranks may be found for sets of raw scores or for scores that have been grouped into intervals in a frequency distribution. In order to provide a single, consistent computational approach, it is always possible to form a grouped-data distribution with an interval width of 1 unit when raw scores are involved and to utilize the grouped-data approach. Thus, the scores 28, 25, 25, 24, 19, 17, 16, 16, 16, 14, 11, 9 can be represented in a frequency distribution as shown in the table below.

Score	True limits	Frequency
28	27.5–28.5	1
25	24.5–25.5	2
24	23.5–24.5	1
19	18.5–19.5	1
17	16.5–17.5	1
16	15.5–16.5	3
14	13.5–14.5	1
11	10.5–11.5	1
9	8.5–9.5	1
		12

Whether we start with a grouped-data distribution or with a frequency distribution of raw scores, the computational procedure is the same. We start by cumulating frequencies from the lowest to the highest interval, converting these to proportions, and then performing an appropriate linear interpolation to determine the desired percentile or percentile rank. To exemplify this process, the 55 quiz scores for which a frequency distribution was previously built (Table 4.3) will be used again. Table 4.9 shows the original frequency distribution with columns added for "cumulative frequency" and "cumulative proportion." The cumulative frequencies are found by determining the total number of scores *at or below* a given score interval. This is most easily done by starting at the interval for the smallest scores; the cumulative frequency for this interval is always the same as the frequency for the interval. The cumulative frequency for the next interval may be found by adding its frequency to the just previous cumulative frequency. Repeating this process will generate cumulative frequencies for all intervals. In the example (Table 4.9), the first cumulative frequency is 3, since this is the frequency for the first interval. To find the cumulative frequency for the second interval, we add its frequency, 1, to the previous cumulative frequency, 3, to obtain 4.

TABLE 4.9   Cumulative percentage distribution for 55 quiz scores

Scores	True limits	Frequency	Cumulative frequency	Cumulative proportion
38–40	37.5–40.5	14	55	1.0000
35–37	34.5–37.5	11	41	.7455
32–34	31.5–34.5	9	30	.5455
29–31	28.5–31.5	7	21	.3818
26–28	25.5–28.5	5	14	.2545
23–25	22.5–25.5	4	9	.1636
20–22	19.5–22.5	1	5	.0909
17–19	16.5–19.5	1	4	.0727
14–16	13.5–16.5	3	3	.0545

The values in the "cumulative proportion" column were found by converting each cumulative frequency to a proportion of 55. Thus, the cumulative proportion for the first interval is 3/55, or .0545; for the interval 29–31, the cumulative frequency is 21 and the cumulative proportion is 21/55, or .3818.[†]

[†] At least four decimal places should be computed to ensure the accuracy of additional computations based on these values.

Finding cumulative proportions is greatly facilitated—especially if calculating machines are being used—by the following procedure: let N be the total number of scores in the table; compute the quantity $1/N$ and express it as a decimal number; and *multiply* each cumulative frequency by $1/N$ to obtain the corresponding cumulative proportions. For example, $1/N = \frac{1}{55}$, or .01818; for the first interval, $3(.01818) = .0545$; for the second interval, $4(.01818) = .0727$; and so forth.

By means of the "cumulative proportion" and "true limits" columns, any desired percentile or percentile rank can be found. First the procedures for finding percentiles will be considered. In outline, any percentile can be found by inspecting the table to locate the score interval containing the percentile. Then, by use of linear interpolation one can find the score value that cuts off the desired proportion of scores. If the 40th percentile is desired for the data in Table 4.9, one must start by inspecting the "cumulative proportion" column to locate the interval containing P_{40}. This is the interval with true limits 31.5 and 34.5, since .3818 of the scores are below 31.5 and .5455 are below 34.5; hence, some score point between 31.5 and 34.5 must have 40 percent of the scores below it. Linear interpolation within the interval is carried out by locating the score value in the interval 31.5–34.5 at a distance proportional to the distance .40 is in the interval .3818–.5455. The proportion .40 is .0182/.1637 of the way from .3818 to .5455. Thus, P_{40} is .0182/.1637 of the way from 31.5–34.5. Then, $P_{40} = 31.5 + (.0182/.1637)(3) = 31.5 + .33$, or 31.83.

As a further example, consider finding P_{90}, the score value below which 90 percent of the scores fall. The 90th percentile falls in the interval 37.5–40.5. The proportion, .90, is .1545/.2545 of the way from 27.5 to 40.5; thus, $P_{90} = 37.5 + (.1545/.2545)(3)$, or 39.32.

The logic underlying this procedure is that in the absence of specific information concerning the scores in an interval, we assume that they are evenly spread out over the interval so as to completely fill it. The interval with true limits 22.5–25.5, for example, contains 4 scores. We assume that these are spread out as follows:

```
        1        1        1        1
   | score | score | score | score |
22.5                              25.5
```

so that the entire interval is filled. When linear interpolation is performed to find a percentile, we move a certain proportion of the way into the interval and may include some fractional part of a score below a percentile. To clarify this point further, let us consider what is obtained when we compute P_{90}. The 90th percentile has 90 percent of 55, or 49.5, scores below it. Thus, it is neces-

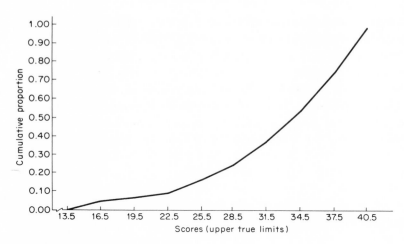

figure 4.6   *Ogive for 55 quiz scores.*

sary in computing P_{90} to include a fractional part, $\frac{1}{2}$, of a score below P_{90}.
The assumption that the scores are spread out in the interval provides a basis
for including just part of a score below any particular percentile we may be
computing.

The problem of determining the percentile rank of a score is also solved by
linear interpolation. In this case, interpolation is carried out in the "cumulative proportion" column. As an illustration, the percentile rank for the score
21 will be found from Table 4.9. The score 21 is $(1.5)/3$ or $\frac{1}{2}$ of the way
through the interval 19.5–22.5. Since .0727 of the scores are below 19.5 and
.0909 are below 22.5, by interpolation $.0727 + 1/2(.0182) = .08$. Thus, 21
has the percentile rank 8.

Whereas this procedure for computing percentiles and percentile ranks is
satisfactory when only a few such values are desired, it is somewhat simpler to
use a graphic procedure when large numbers of percentiles and/or percentile
ranks must be found. This graph is obtained by plotting the cumulative-pro-
portion values against the upper true limits of the intervals and connecting
adjacent pairs of points by straight-line segments. The resulting graph will
usually appear to be S shaped and is known as an *ogive*. Figure 4.6 displays
the ogive for the data in Table 4.9. Percentiles and percentile ranks can be
read rapidly from the ogive by the usual procedures for reading line graphs.
In order to obtain reasonably accurate values, the ogive should be plotted on as
large a scale as possible. It should be noted that reading a value from the
ogive is really equivalent to the linear-interpolation procedure used with frequency-distribution tables, since the successive points along the graph are connected by straight-line segments.

Certain percentiles have special names, since they divide up the score distribution into equal parts. The three values, P_{25}, P_{50}, and P_{75}, divide the distribution into quarters; hence, they are known as the quartiles and are often denoted Q_1, Q_2, and Q_3, respectively. From Fig. 4.5, Q_1 is found to be 28.4, Q_2 to be 33.7, and Q_3 to be 37.6. These three values divide the distribution into quarters. Scores below 28.4 are in the bottom quarter, scores between 28.4 and 33.7 are in the second quarter, and so forth. Similar to the quartiles are the deciles, P_{10}, P_{20}, . . . , P_{90}, which divide the distribution into 10 equal parts.

4.3 Averages

This is the first of three sections devoted to a discussion of derived measures used to describe and summarize characteristics of score distributions. The tabular and graphic techniques presented in the preceding section are useful devices for communicating large numbers of scores to others and for presenting a "picture" of a score distribution. The derived measures to be presented now allow the researcher to go beyond a general picture of his data and to make quantitative statements concerning specific aspects of the distribution of score with which he is working.

In order to facilitate presentation of the concepts developed in the next three sections, the frequency polygon is used as the basic tool for summarizing score distributions. Moreover, in many cases, an "idealized" representation of a frequency polygon will be used in place of a more accurate and complete figure. The reason for this usage is that we shall want to focus on specific characteristics of score distributions without being concerned with finer details. For example, in this section we shall deal with a class of derived measures known as *averages*. An average is a single number that indicates the position of a distribution of scores along a score axis. The "idealized" frequency polygons in Fig. 4.7 show three score distributions, I, II, and III, which are positioned at different locations along the score axis. Frequency polygon I is concentrated near the low end of the score axis; that is, the scores in the distribution are relatively small values. Frequency polygon III is concentrated at the high end of the axis; and II is situated at an intermediate position. In addition to differing in location, these three frequency polygons also illustrate some of the characteristics discussed in the next two sections. First, they exhibit difference in variability, or spread, of scores. This is most apparent if frequency polygons II and III are compared; the scores in frequency polygon III tend to occur over only a small portion of the score axis, whereas the scores in frequency polygon

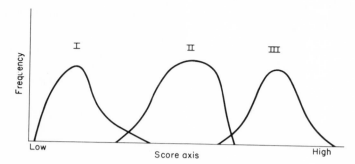

figure 4.7   Frequency polygons differing in location.

II are considerably more disperse.   Derived measures describing the degree of compactness or dispersion of a set of scores are classified as *measures of variability*. Second, the frequency polygons display degrees of symmetry or lack of symmetry. Frequency polygon III is perfectly symmetrical: the right and left halves of the figure are mirror images of one another.   Frequency polygon II displays some lack of this kind of symmetry, and I is markedly asymmetrical.   A figure that lacks symmetry is described as *skewed;* and measures of skewness are available.

Let us now turn to averages—the major topic of this section.   Each average has the purpose of supplying a single number that is, in some sense, "most typical" of all of the scores in a distribution.   In statistics, there are many different kinds of averages, the most important of which are discussed in this section.   The fact that there is more than one available average reflects the fact that the phrase "most typical" can be interpreted in more than one way.   There are, in current usage, three major ways in which a most-typical value for a score distribution is derived.   These will be considered in turn.

(1) The score value that occurs most frequently in a set of scores is one interpretation of a most-typical value.   This score is referred to as the *mode* of the distribution of scores and is symbolized M.   Thus, for the 10 scores 21, 18, 17, 20, 21, 14, 17, 19, 17, 20, there is one score that occurs more frequently than any other; this score is 17, and therefore $M = 17$.   Problems that arise when one attempts to locate the mode of a set of scores are either that no score occurs more frequently than the others (e.g., the set of scores, 14, 18, 20, 20, 14, 18 has no mode because each occurs exactly twice) or that two or more scores are tied for occurring most often (e.g., the set of scores 21, 14, 19, 21, 19, 11, 16, 19, 13, 21 has two modes—one at 19 and one at 21).   In the first situation, we are faced with the undesirable fact that no mode can be determined from the data; an average, if it is to be widely used, should always be defined on a set of data.   In the second situation, we may end up with two, three, or more modes for one set of scores.   This is not necessarily undesirable, since the existence of

multiple modes tells us a good deal about the shape of the score distribution. Technically, distributions with two modes are described as *bimodal*, those with three modes as *trimodal*; and, in general, the term multimodal may be used to refer to distributions with more than three modes.

When data are summarized in the form of a frequency-distribution table, the mode is defined as the class index of the score interval with the largest frequency. In the frequency distribution in Table 4.9, the last category, 38–40, has the largest frequency. Hence, the mode, or 39, is the midpoint of this interval. Of course, frequency distributions may also be bimodal, trimodal, or multimodal.

The concept of the mode is often extended to include the occurrence of *local modes*. To make a clear distinction, the most frequently occurring score, or class index of the score interval with maximum frequency, is called the absolute mode. A local mode exists when a score, or score interval, occurs more frequently than its immediate neighbors on both sides but is still not the most frequent in the entire distribution. Consider the frequency distribution in Table 4.10 and the frequency polygon based on it (Fig. 4.8). The absolute mode is clearly 72 because the interval 70–74 has the maximum frequency. However, a well-defined local mode exists at 27 (i.e., in the interval 25–29), as

TABLE 4.10   Frequency distribution for 150 scores

Score interval	Frequency
85–89	2
80–84	5
75–79	15
70–74	24
65–69	18
60–64	11
55–59	6
50–54	2
45–49	2
40–44	6
35–39	9
30–34	14
25–29	16
20–24	12
15–19	5
10–14	3
Total	150

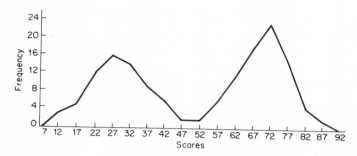

figure 4.8   *Frequency polygons for 150 scores.*

is apparent from inspecting the frequency polygon. In practice, the occurrence of more than one mode (either two or more modes, or one absolute mode and one or more local modes) should alert the researcher to the possibility that scores in fact have been combined for two or more distinctive populations or subpopulations. Suppose that the data summarized in Fig. 4.8 were scores on a test of arithmetic-computation skill for four classes of ninth graders taught by one teacher. One explanation for the existence of two modes might be that the classes were each homogeneously grouped on the basis of previous mathematics achievement and that two of the classes were quite low groups and two were high groups. In effect, two distinct subpopulations, the low and the high mathematics groups, exist among the 150 students for whom scores are available.

(2) A second interpretation of a most-typical value for a score distribution is the score point that is in the middle of the distribution. This average is known as the *median;* it is the score value above which half of the scores lie. Note that the median is equivalent to the 50th percentile, since, by definition, P_{50} has 50 percent of the scores at or below it. The usual symbol for the median is then P_{50}.

When data are grouped into score intervals to form a frequency distribution, the median may be computed by the same procedure as any other percentile. That is, the technique involving linear interpolation between the true limits of the score containing P_{50}, which was described in the previous section of this chapter, may be used. When computation is carried out with original scores (i.e., with ungrouped data), certain factors must be considered. Simple counting procedures may fail to locate a median for a set of scores. For example, the scores 83, 83, 81, 78, 39 do contain a value that lies in the middle; this is the score 81 and it is the median. On the other hand, the scores 42, 41, 36, 32 contain no middle score. In fact, whenever an even number of scores is being considered, there will be no middle score. For this case, the score point halfway between the *two* middle scores is taken to be the median. Thus, 38.5 is the median of the four scores because it lies exactly halfway between 36 and 41.

Whether there is an even number or an odd number of scores, it may still happen that the value of the median is not immediately apparent. The scores 26, 26, 25, 22, 22, 22, 19, although odd in number, nevertheless have no middle score. This is so because 22 recurs 3 times near the middle of the distribution. The score 22 cannot be taken as the median because only one score is below 22, while 3 scores are above 22. The occurrence of tied scores as well as the minor problem when an even number of scores is present suggests that some systematic procedure, not based entirely on counting, is necessary to give a consistent definition to the median. Such a procedure is available and operates as follows: whenever it is necessary to compute a median from ungrouped data, treat the score distribution as if it were a grouped-data frequency distribution with an interval width equal to the unit of measurement for the scores; then compute P_{50} using the procedure for frequency distributions. For most educational data, the unit of measurement is a whole number. This procedure should be used with all sets of original scores and it easily generalizes to the computation of percentiles other than the 50th when computation is based on original scores. Let us consider an example. The scores in Table 4.11

TABLE 4.11 Scores for 15 students

23	8	15
8	20	18
14	12	14
9	12	19
11	16	9

are shown in the form of a whole-unit frequency distribution in Table 4.12 (note that nonoccurring score values, that is, those with 0 frequency, are omitted from the table). In Table 4.12 the "true limits," "cumulative frequency," and "cumulative proportion" columns have been added because these are required for computing any percentile. Note that the median, P_{50}, falls between 13.5 and 14.5, since .4667 of the scores are below 13.5 and .6000 of the scores are below 14.5. Using linear interpolation, we locate a value that is .0333/.1333 of the way from 13.5 to 14.5. This is 13.5 + (.0333/.1333) = 13.5 + .25, or $P_{50} = 13.75$.

The rationale underlying this computing procedure is the same as that for the computation of percentiles from a grouped-data frequency distribution. That is, we conceive of the scores in a score interval as evenly spread out over

TABLE 4.12   *Whole-number frequency distribution*
for 15 scores

Score	True limits	Frequency	Cumulative frequency	Cumulative proportion
23	22.5–23.5	1	15	1.0000
20	19.5–20.5	1	14	.9333
19	18.5–19.5	1	13	.8667
18	17.5–18.5	1	12	.8000
16	15.5–16.5	1	11	.7333
15	14.5–15.5	1	10	.6667
14	13.5–14.5	2	9	.6000
12	11.5–12.5	2	7	.4667
11	10.5–11.5	1	5	.3333
9	8.5–9.5	2	4	.2667
8	7.5–8.5	2	2	.1333
		15		

the entire interval so as to completely fill the interval from one true limit to the other. Then, a score value is determined that has exactly the required percentage of cases below it. This often turns out to be a decimal number even though the original scores are whole numbers.

(3) A third interpretation of the phrase most-typical score is the average that is nearest to all scores in a distribution in the sense that deviations from this value are minimal. This measure is the *arithmetic mean* (symbolized μ), which is found by summing the scores and then dividing this sum by the number of scores. If we let X_i be the ith score in a distribution of N scores, then the arithmetic mean μ is

$$\frac{\sum_{i=1}^{N} X_i}{N}$$

That the deviations from the arithmetic mean are minimal can be shown as follows: the deviation of any individual score X_i from the arithmetic mean is the quantity $(X_i - \mu)$.† The sum of the deviations of all scores in a distribution from μ is, then,

$$\sum_{i=1}^{N} (X_i - \mu)$$

† This deviation can be positive or negative. If the absolute deviation is defined as $|X_i - \mu|$, then it is *not* true that the arithmetic mean has minimal absolute deviation. As a matter of fact, the quantities $|X_i - P_{50}|$ have this property.

It can be shown that

$$\sum_{i=1}^{N} (X_i - \mu) = 0$$

First, remove the parentheses and apply the summation operator to each term:

$$\sum_{i=1}^{N} (X_i - \mu) = \sum_{i=1}^{N} X_i - \sum_{i=1}^{N} \mu = \sum_{i=1}^{N} X_i - N\mu$$

But

$$\mu = \frac{\displaystyle\sum_{i=1}^{N} X_i}{N}$$

hence

$$\sum_{i=1}^{N} (X_i - \mu) = \sum_{i=1}^{N} X_i - N \frac{\displaystyle\sum_{i=1}^{N} X_i}{N} = \sum_{i=1}^{N} X_i - \sum_{i=1}^{N} X_i = 0$$

Also, the only score value which has the property that deviations from it sum to 0 is the arithmetic mean. This is evident if we let c be any other score value; then, let $d = c - \mu$ be the difference between c and μ. Rearranging terms, this becomes $c = \mu + d$. The quantity

$$\sum_{i=1}^{N} (X_i - c)$$

is equivalent to

$$\sum_{i=1}^{N} [X_i - (\mu + d)]$$

Then

$$\sum_{i=1}^{N} [X_i - (\mu + d)] = \sum_{i=1}^{N} X_i - N\mu - Nd = \sum_{i=1}^{N} X_i - N \frac{\displaystyle\sum_{i=1}^{N} X_i}{N} - Nd$$

$$= \sum_{i=1}^{N} X_i - \sum_{i=1}^{N} X_i - Nd = -Nd$$

If d is any value other than 0, we see that

$$\sum_{i=1}^{N} (X_i - c)$$

is different from 0.

Computation of the arithmetic mean from a set of original scores is a familiar process to most readers of this text. To compute the arithmetic mean for the scores in Table 4.13, we sum all scores to find

$$\sum_{i=1}^{15} X_i = 1,253$$

Then, since there are 15 scores, $\mu = 1253/15 = 83.53$. Computation of the arithmetic mean can also be performed from a frequency distribution. Since this procedure is greatly simplified by using a technique known as "coding," consideration is deferred until the last section of this chapter.

TABLE 4.13   Scores for 15 students

81	87	93
91	64	86
58	72	74
96	90	90
100	84	87

The three averages described in this section, the mode, the median, and the arithmetic mean, are the most commonly used of the available measures. No other average will be required for any of the procedures developed in the remainder of this book. However, the geometric mean and the harmonic mean are worthy of mention at this point because the student will occasionally come into contact with them. The computation of the geometric mean is similar in form to that of the arithmetic mean with multiplication replacing addition and with the extraction of a root replacing division. That is, the geometric mean, μ_g, of N scores is $\mu_g = \sqrt[N]{X_1 \cdot X_2 \cdot X_3 \cdots X_N}$, the Nth root of the product of the N scores. This can be written in more compact notation using the multiplication operator

$$\mu_g = \left[\prod_{i=1}^{N} X_i \right]^{1/N}$$

Two features of the geometric mean should be noted. (1) If any one of the X_i is 0, then μ_g is 0; also, if an odd number of scores is negative, then μ_g is an imaginary number. (2) If the scores X_i are replaced by their logarithms, log X_i, then the arithmetic mean of these logarithmic values is the logarithm of μ_g; this suggests a simplified computing procedure for μ_g by means of logarithms, since the product of N scores is a rather unwieldy quantity.

The harmonic mean, denoted μ_h, is the reciprocal of the arithmetic mean of the score reciprocals. That is,

$$\mu_h = \frac{N}{\sum\limits_{i=1}^{N} (1/X_i)}$$

Note that a 0 score results in an indeterminable harmonic mean.

In practical situations, it is not uncommon for the researcher to be faced with a decision concerning which of the averages is most appropriate to report for a given type of data with which he is working. We shall limit our consideration to the mode, median, and arithmetic mean because these are appropriate to the vast majority of practical problems. Whereas a large number of criteria could be listed along which these measures might be compared, we shall limit ourselves to just five requirements. (1) Ease of computation. When calculating machines are available, this criterion becomes trivial; it is mentioned here only because some individuals are prone to select measures on this basis, which we consider irrelevant in almost any situation in educational research; thus, the first criterion is a pseudo criterion; (2) Universality of definition. This criterion refers to whether or not there are score distributions for which the average cannot be computed. Whereas the median and arithmetic mean can be found for any set of scores, the mode is undefined for a distribution of original scores if all scores occur only once, or the same number of times, and for a grouped-data distribution if each interval has the same frequency. (3) Sensitivity to deviant scores. Some score distributions have a few scores that are located far from the bulk of the scores. When these deviant scores occur in just one direction, either toward the high end or the low end of the score dimension, the arithmetic mean is more affected than is either the median or the mode. For example, consider the 10 scores, 21, 22, 23, 23, 23, 24, 25, 28, 50. The deviant score is 50; the mode is 23; the median is 23.3; and the arithmetic mean is 26.6. Note that the one high score has made the arithmetic mean much larger than the median; that is, the one large score has "pulled" the arithmetic mean toward the high end of the score dimension. Also, if no deviant score had occurred, i.e., if the highest score had been 29 or 30 instead of 50, the values of the mode and the median would be completely unchanged, although the arithmetic mean would have a smaller value. As was mentioned earlier in this section, score distributions with extreme scores in one direction are described as "skewed" distributions. Hence, we may state that the arithmetic mean is more affected by the presence of skewness than is the mode or median. This is an important consideration when selecting measures of location for actual score distributions. A good example involves the case of family income data for defined groups of individuals. If a city, county, state, or other political division announces "average" income figures for families in their area, it is necessary to know which average is being reported. A few wealthy families in an area may raise the average considerably if the arithmetic mean is the reported value. In fairness, it should be noted that in almost all cases involving data of this type, it is the median that is reported. (4) Total score dependence. This criterion refers to the degree to which the average is based on the actual scores in the distribution. The mode and median do not depend upon the actual numerical values of all scores in a distribution. The arithmetic mean, on the other hand, uses these numerical values directly in its computation and can be said to depend upon each score in the distribution. (5) Use in advanced application. It is on this criterion that the clearest distinctions exist among the three averages. The mode rarely appears in use beyond the simplest descriptive applications. Whereas there exists a body of advanced theory related to the median, little of this is in use in the analysis of data in applied-research situations. The arithmetic mean, on the other hand, is extensively employed in advanced applications, and a vast amount of theory has been built up around the arithmetic mean. The basic reason for these contrasts among the three averages is the comparative ease with which algebraic and higher-level mathematical manipulations may be carried out with the measures.

In summary, the arithmetic mean is most often the value that is chosen to be reported as the average. This choice is based on the fact that it can always be computed and, more basically, because of its central role in advanced theory and applications of statistics. However, for descriptive purposes, the median is often more useful than the arithmetic mean when skewed distributions are involved.

4.4 Measures of variability

A second major characteristic of score distributions is their spread, or variability, along the score dimension. As was the case for averages, there is a variety of measures of variability available to the researcher. Each of these is usually associated with one of the averages. That is, for each average, the mode, median, and arithmetic mean, there exists one or more measures of variability more appropriate than others to report along with that average. We shall consider the measures of variability in the same order of presentation as the corresponding average.

The range w of a set of scores, which has been previously discussed, is the number of score units covered by the scores. A general formula is $w = X_{max} - X_{min} + 1$ unit, where X_{max} is the largest score in the distribution,

TABLE 4.14   Scores on a problem assignment for 12 statistics students

Score	$\|X - P_{50}\|$	$\|X - \mu\|$	Score	$\|X - P_{50}\|$	$\|X - \mu\|$
57	5	2.6667	51	11	8.6667
40	22	19.6667	65	3	5.3333
59	3	.6667	66	4	6.3333
69	7	9.3333	69	7	9.3333
63	1	3.3333	62	0	2.3333
53	9	6.6667	62	0	2.3333

X_{min} is the smallest score, and 1 unit is the unit of measurement for the scores. In most educational applications, the unit of measurement is a whole number, so that $w = X_{max} - X_{min} + 1$. An example of the computation of w was given in Sec. 4.2. The range is a rather crude measure of variability, since it does not depend upon any scores except the two most extreme (that is, X_{max} and X_{min}). The relative density or dispersion of scores between these extremes in no way influences the value of the range. In general, the range is reported as a measure of variability when the mode is utilized as the average. For grouped-data distributions, the range may be taken as the difference between the upper true limit of the highest-score interval and the lower true limit of the lowest-score interval.

When the median is chosen as the average, two major possibilities exist as choices for quantifying variability. The first of these, the *average deviation*, is the arithmetic mean of the absolute values of all deviations from the median and is denoted Δ. In symbols,

$$\Delta = \frac{\sum_{i=1}^{N} |X_i - P_{50}|}{N}$$

As was mentioned previously, the value of the numerator of the right side of the equation is minimized when P_{50} is used; that is, if P_{50} is replaced by any other value (including μ)†, then the resulting value will be no smaller and, in general, larger than when P_{50} is used. The computation of Δ will be illustrated for the scores in Table 4.14. The median is 62 and the arithmetic mean is 59.6667.

† Other authors sometimes define the *mean deviation* as

$$\frac{\sum_{i=1}^{N} |X_i - \mu|}{N}$$

This quantity is so rarely used in applied statistics that we omit discussion of it.

The columns $|X - P_{50}|$ and $|X - \mu|$ allow a comparison of

$$\sum_{i=1}^{N} |X_i - P_{50}| = 72 \qquad \text{and} \qquad \sum_{i=1}^{55} |X_i - \mu| = 76.6666$$

Note that the sum of absolute deviations about P_{50} is less than about μ. The average deviation is then $\Delta = 72/12 = 6$. In words: on the average, scores deviate by 6 units, in absolute value, from the median. The average deviation is not a widely used measure and is rarely computed from grouped-data distributions. For this reason, no computing routine is presented here for frequency distributions.

An alternative measure of variability, which is often associated with the median, is the *semi-interquartile range*, denoted Q. The semi-interquartile range is one-half the difference between the 25th and 75th percentiles of the score distribution; that is, $Q = (P_{75} - P_{25})/2$, or $(Q_3 - Q_1)/2$. In contrast to the simple range, Q is not dependent on the two most extreme scores in a distribution; however, it does not use all score values in its computation, as does the average deviation. When interpreting a value of Q, it is important to avoid a common error. If the quantities $P_{50} + Q$ and $P_{50} - Q$ are computed, they are, in general, *not* equal to P_{75} and P_{25}, respectively. In fact, they equal P_{75} and P_{25} only for perfectly symmetric distributions. This can be seen from Fig. 4.9 in which frequency distribution I is symmetric and frequency distribution II is skewed.

For the arithmetic mean, an appropriate measure of variability is the *variance* σ^2 and its square root, the *standard deviation* σ. The variance is the arithmetic mean of the *squared* deviations from the arithmetic mean. In symbols,

$$\sigma^2 = \frac{\sum_{i=1}^{N} (X_i - \mu)^2}{N}$$

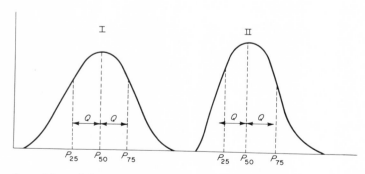

figure 4.9   *Relationship between Q and the quartiles for a symmetric and an asymmetric distribution.*

Note that the deviations, $(X_i - \mu)$, are first squared to find $(X_i - \mu)^2$ and then summed to yield

$$\sum_{i=1}^{N} (X_i - \mu)^2$$

The variance is found by dividing this final quantity by N. The quantity

$$\sum_{i=1}^{N} (X_i - \mu)^2$$

is known as the *sum of squares*. Thus, the variance is the sum of squares divided by the number of scores N. The student should not confuse the sum of squares

$$\sum_{i=1}^{N} (X_i - \mu)^2$$

with the sum of squared scores

$$\sum_{i=1}^{N} X_i^2$$

Since the deviations, $(X_i - \mu)$, are squared during the computation of the variance, the resulting value for σ^2 is in squared units. That is, if the X_i are IQ scores, then σ^2 is squared IQ units. Since it is difficult to conceptualize the meaning of a squared IQ unit, or squared units along many other educational dimensions, the square root of σ^2, or the standard deviation σ, is ordinarily used for reporting purposes. The standard deviation will be in IQ units if the original scores are IQ scores. The complete formula for the standard deviation is

$$\sigma = \sqrt{\frac{\sum_{i=1}^{N} (X_i - \mu)^2}{N}}$$

This formula will be referred to as the "definitional formula" for σ. It has disadvantages with respect to computation. First, the value of μ is usually a decimal number, which means that the $(X_i - \mu)$ will be decimal numbers; thus, finding $(X_i - \mu)^2$ involves squaring decimal numbers. Second, a good deal of subtraction must be carried out to find the $(X_i - \mu)$ values. This is inconvenient by hand or on a calculating machine. A simpler computational formula can be derived. This will be done for the variance in order to avoid repetitious writing of the square-root radical. The numerator can be expanded to yield

$$\frac{\sum_{i=1}^{N} (X_i - \mu)^2}{N} = \frac{\sum_{i=1}^{N} (X_i^2 - 2\mu X_i + \mu^2)}{N}$$

The summation operator is then applied:

$$\frac{\sum\limits_{i=1}^{N}(X_i^2 - 2\mu X_i + \mu^2)}{N} = \frac{\sum\limits_{i=1}^{N}X_i^2 - 2\mu\sum\limits_{i=1}^{N}X_i + N\mu^2}{N}$$

Substituting

$$\mu = \frac{\sum\limits_{i=1}^{N}X_i}{N}$$

then

$$\sigma^2 = \frac{\{\sum\limits_{i=1}^{N}X_i^2 - 2\left[(\sum\limits_{i=1}^{N}X_i)/N\right](\sum\limits_{i=1}^{N}X_i) + N\left[(\sum\limits_{i=1}^{N}X_i)/N\right]^2\}}{N}$$

$$= \frac{\sum\limits_{i=1}^{N}X_i^2 - 2(\sum\limits_{i=1}^{N}X_i)^2/N + (\sum\limits_{i=1}^{N}X_i)^2/N}{N}$$

$$= \frac{\sum\limits_{i=1}^{N}X_i^2 - (\sum\limits_{i=1}^{N}X_i)^2/N}{N}$$

This result may be written in several equivalent forms: e.g.,

$$\sigma^2 = \frac{\sum\limits_{i=1}^{N}X_i^2 - (\sum\limits_{i=1}^{N}X_i)^2/N}{N} \quad \text{or} \quad \sigma^2 = \frac{\sum\limits_{i=1}^{N}X_i^2 - N\mu^2}{N}$$

$$\text{or} \quad \sigma^2 = \frac{\sum\limits_{i=1}^{N}X_i^2}{N} - \mu^2$$

The first and second formulas will be used in this text. Note that using them involves finding

$$\sum_{i=1}^{N}X_i^2$$

the sum of the squared scores and

$$\sum_{i=1}^{N}X_i$$

the sum of the scores; both of these quantities can be accumulated simultaneously when a calculating machine is used.

TABLE 4.15   Scores and squared scores for 12 students on a statistics-problem assignment

X	X^2	X	X^2
57	3,249	51	2,601
40	1,600	65	4,225
59	3,481	66	4,356
69	4,761	69	4,761
63	3,969	62	3,844
53	2,809	62	3,844
		716	43,500

The computational formula for σ^2 will be used to find the variance of the 12 scores in Table 4.14. These are reproduced in Table 4.15 along with the X_i^2 values. The sum of the X column is

$$\sum_{i=1}^{12} X_i = 716$$

and the sum of the X^2 column is

$$\sum_{i=1}^{12} X_i^2 = 43,500$$

The arithmetic mean is, then,

$$\mu = \frac{\sum_{i=1}^{N} X_i}{N} = \frac{716}{12} = 59.6667$$

The variance is

$$\sigma^2 = \frac{43,500 - (716)^2/12}{12} = \frac{43,500 - (512,656/12)}{12}$$

$$= \frac{43,500 - 42,721.3333}{12} = \frac{778.6667}{12} = 64.8889$$

The standard deviation is the square root of the variance; thus,

$$\sigma = \sqrt{64.8889} = 8.0554$$

A procedure for computing the variance from a grouped-data frequency distribution will be presented in Sec. 4.6.

The measures of variability considered in this section can be compared against the same criteria previously applied to averages: (1) Ease of computation: The range is certainly the easiest measure to compute; however, with modern computing aids, especially the desk calculating machine and the high-speed digital computers, this is a trivial point to ponder. (2) Universality of definition: Each of the measures of variability satisfies this criterion. (3) Sensitivity to deviant scores: The range is obviously extremely sensitive to deviant scores. A single score value far removed from the bulk of the scores will almost completely determine the value of w; among the remaining measures, the variance is more sensitive to extreme scores than is, say, the average deviation or semi-interquartile range. However, in the case of a measure of variability, this sensitivity to extreme scores is not necessarily a weakness, since it is variability which we are quantifying. (4) Total score dependence: The range and the semi-interquartile range both fall down on this criterion, since they each may depend only upon two scores in the distribution. The average deviation and variance (and standard deviation) do depend upon all scores. (5) Use in advanced applications: On this criterion, the variance and its square root, the standard deviation, are far ahead of the other measures of variability. Since the variance is associated with the arithmetic mean, the previous remarks concerning the arithmetic mean apply to the variance as well.

In summary, it is the arithmetic mean as the average and the variance as a measure of variability that play a central role in the more advanced theory and practice in applied statistics. For this reason, these two measures will receive a good deal of treatment in the remainder of this textbook.

4.5   Other characteristics of score distributions

The location of a score distribution along the score axis and its variability are major features that will be of primary concern in future chapters. In addition to these characteristics, the notions of skewness and of relative peakedness or flatness of a distribution are important descriptive features in many situations. The concept of skewness has been introduced previously in this chapter; it refers to the degree to which a distribution lacks symmetry. Two distributions may be identical in location and in variability but yet differ in symmetry. Figure 4.10 illustrates how this might occur. Note that distribution I and II are each asymmetric and have the same arithmetic mean but different lack of symmetry.

In describing skewed distributions, special terms are used to refer to the manner in which a distribution lacks symmetry. A frequency distribution such as I in Figure 4.10 is described as *skewed to the right*, or *positively skewed*,

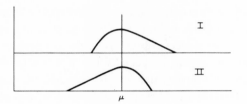

figure 4.10   *Two distributions skewed in opposite directions.*

since the figure has a "tail" toward the right; frequency polygon II is described as *skewed to the left*, or *negatively skewed*, since its "tail" is to the left.   At this point, these terms are rather vaguely defined, since a distribution may lack symmetry yet not have a readily apparent "tail" to the right or to the left.   In order to quantify degree of skewness, a commonly used measure is the skewness index $S_k = 3(\mu - P_{50})/\sigma$.   The sign of the skewness index determines whether a distribution is positively or negatively skewed.   Note that the sign of S_k depends upon the difference $\mu - P_{50}$.   If the arithmetic mean is larger than the median, the sign of $\mu - P_{50}$ is positive and the distribution is positively skewed; if P_{50} is larger than μ, $\mu - P_{50}$ is negative in sign, and the distribution is negatively skewed.   This is consistent with the previous discussion of tails on a score distribution.   For a distribution with a tail to the right, the arithmetic mean will be pulled toward that tail (i.e., toward the high end of the score dimension) and μ will be larger than P_{50}; if the tail is toward the left, then μ is pulled toward the low-score end of the dimension and μ will be smaller than P_{50}.   In a symmetric distribution the arithmetic mean and the median will, of course, be identical in value and $\mu - P_{50}$ will be 0 (as will S_k).   Figure 4.11 illustrates these situations.

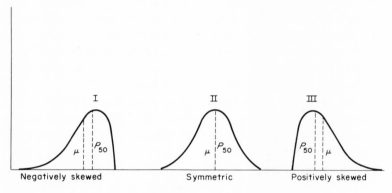

figure 4.11   *Three distributions differing in skewness.*

The degree to which a distribution is relatively peaked or relatively flat is described as *kurtosis*.   In comparing two or more distributions, the following language is used: a relatively flat distribution is described as *platykurtic* (see frequency polygon I, Fig. 4.12), a moderately peaked distribution is described as *mesokurtic* (see frequency polygon II, Fig. 4.12), and a relatively peaked distribution is described as *leptokurtic* (see frequency polygon III, Fig. 4.12).

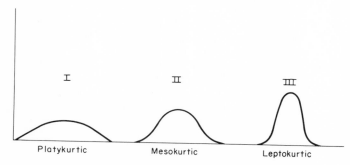

figure 4.12   Three distributions differing in kurtosis.

No measure of kurtosis is presented because none is in general use in educational research.   The concept of kurtosis is useful for a simple description of score distributions but is rarely encountered in advanced applications.

4.6   Coding of data

The term "coding" is used to refer to any process of transforming scores from their original state into some new score system.   A simple example of coding arises when some arbitrary constant is subtracted from each score in a distribution in order to make the resulting numbers easier to work with (e.g., if scores are IQs and each is larger than 100, it might be convenient to subtract 100 from each score so that computations can be done with 2-digit instead of 3-digit numbers).   In this section, we shall consider certain types of coding transformations and their effects upon the resulting data analysis.

The class of transformations considered here is limited to so-called "linear" transformations.   That is, if X_i is an original score and X_i' is a coded score, then X_i is transformed into X_i' by an equation of the form $X_i' = c + dX_i$, where c and d are any two arbitrary constants.   Note that this equation is of the general class of equations that represent straight lines when plotted on Cartesian coordinates.   Thus, if we assign specific numerical values to c and d, the resulting equation is a straight line.   Hence, scores coded in this way are linearly related to the original scores. It should be noted that c or d can be *any* numbers, positive, negative, fractional, etc.

If data are coded, then the arithmetic mean and variance of the coded scores bear simple relationships to the arithmetic mean and variance of the original scores. The arithmetic mean of the X_i values is

$$\mu = \frac{\sum_{i=1}^{N} X_i}{N}$$

and the variance is

$$\sigma^2 = \frac{\sum_{i=1}^{N} (X_i - \mu)^2}{N}$$

For the coded scores,

$$\mu' = \frac{\sum_{i=1}^{N} X_i'}{N} \quad \text{and} \quad \sigma'^2 = \frac{\sum_{i=1}^{N} (X_i' - \mu')^2}{N}$$

If $X_i' = c + dX_i$, then

$$\mu' = \frac{\sum_{i=1}^{N} (c + dX_i)}{N} = \frac{Nc + d \sum_{i=1}^{N} X_i}{N} = c + d\left(\frac{\sum_{i=1}^{N} X_i}{N}\right) = c + d\mu$$

Thus, μ' bears the same relationship to μ as X_i' does to X_i. Also,

$$\sigma'^2 = \frac{\sum_{i=1}^{N} [(c + dX_i) - (c + d\mu)]^2}{N} = \frac{\sum_{i=1}^{N} (dX_i - d\mu)^2}{N}$$

$$= \frac{\sum_{i=1}^{N} d^2(X_i - \mu)^2}{N} = \frac{d^2 \sum_{i=1}^{N} (X_i - \mu)^2}{N}$$

$$= d^2\sigma^2$$

thus, $\sigma'^2 = d^2\sigma^2$. The value of the additive constant c has no effect on σ'^2.

Some special examples of this general result are of interest. If we merely add or subtract a constant from each score, the transformation becomes $X_i' = c + X_i$. This implies that $d = 1$. For this special case, $\mu' = c + \mu$ and $\sigma'^2 = \sigma^2$. That is, the arithmetic mean of the X_i' is c units from the original arithmetic mean μ. However, the variances of the original and coded scores are the same. This apparently surprising result is understandable if Fig. 4.13

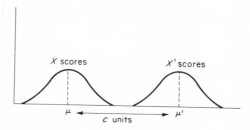

figure 4.13   Original and coded scores, c
                 positive.

is studied. The original distribution of scores is simply moved up along the
score axis c units (assuming that c is a positive number) when the transformation
$X_i' = c + X_i$ is made. The variability of the scores has not been affected;
hence, $\sigma'^2 = \sigma^2$. However, the new arithmetic mean μ' is c units above μ.

To illustrate this outcome in numerical form, the scores in Table 4.16 will
be used. For the original X scores,

$$\mu = \frac{441}{4} = 110.25 \quad \text{and} \quad \sigma^2 = \frac{48{,}985 - (441)^2/4}{4} = 91.19$$

after 100 is subtracted from each score, the new arithmetic mean is

$$\mu' = \frac{41}{4} = 10.25$$

and the variance is

$$\sigma'^2 = \frac{785 - (41)^2/4}{4} = 91.19$$

Thus, $\mu' = \mu - 100$ and $\sigma'^2 = \sigma^2$.

If each score is multiplied (or divided) by a constant and a second constant
is added to (or subtracted from) the result, then $\mu' = c + d\mu$ and $\sigma'^2 = d^2\sigma^2$.

TABLE 4.16   Original and coded scores,
$c = -100$

X	X^2	X'	X'^2
118	13,924	18	324
121	14,641	21	441
98	9,604	−2	4
104	10,816	4	16
441	48,985	41	785

The scores in Table 4.17 illustrate this case. The original arithmetic mean and variance are

$$\mu = \frac{50}{4} = 12.5 \quad \text{and} \quad \sigma^2 = \frac{750 - (50)^2/4}{4} = 31.25$$

For the coded scores,

$$\mu' = \frac{2}{4} = .5 \quad \text{and} \quad \sigma'^2 = \frac{6 - (2)^2/4}{4} = 1.25$$

Then, $\mu' = (1/5)\mu - 2 = 12.5/5 - 2 = 2.5 - 2 = .5$ and $\sigma'^2 = (1/5)^2 \cdot \sigma^2 = 31.25/25 = 1.25$ and these agree with the computed values.

TABLE 4.17 Original and coded scores, $c = -2, d = \frac{1}{5}$

X	X^2	X'	X'^2
20	400	2	4
15	225	1	1
10	100	0	0
5	25	−1	1
50	750	2	6

The technique of coding can be utilized during the computation of the arithmetic mean or variance from a frequency-distribution table. When data are grouped into intervals to form a frequency distribution, computation of the arithmetic mean is somewhat complicated by the fact that we no longer know the values of the individual scores in any one score interval. In order to overcome this problem, it is assumed that each score in a given interval has as its value the class index for that interval. Since real data rarely conform to this assumption, it should be noted that the arithmetic mean computed from a frequency distribution will differ in value from the arithmetic mean as computed on the original scores from which the frequency distribution was built.

Since the numbers in the "class index" column may be large, hand computation is facilitated by first transforming these to some coded form. The class-index column will always be evenly spaced multiples of i, the interval width (not necessarily whole-number multiples, however). Thus, $d = 1/i$ is selected as the multiplicative constant for performing a transformation on the class indexes. Then, some constant, c, is chosen to further simplify these values. For example, the frequency distribution for 55 quiz scores is reproduced in

Table 4.18. The class indexes are whole-number multiples of 3; the column labeled $X*$ shows the values $X* = (\frac{1}{3})X$. To make X' values small, c is chosen as -10. Then $X' = (1/3)X - 10$, or $X' = X* - 10$.

TABLE 4.18 *Use of coded class indexes to compute the arithmetic mean*

Class index, X	$X*$	X'	f	fX'
39	13	3	14	42
36	12	2	11	22
33	11	1	9	9
30	10	0	7	0
27	9	-1	5	-5
24	8	-2	4	-8
21	7	-3	1	-3
18	6	-4	1	-4
15	5	-5	3	-15
			55	38

The values of μ' and σ'^2 may be found from the frequency distribution and these can then be transformed to μ and σ^2 by the following formulas: $\mu = i\mu' - ic$ and $\sigma^2 = i^2 \sigma'^2$. In computing these quantities, the sign of c must be taken into account.

Computation of μ' from the frequency distribution is more conveniently accomplished by forming a new column on the table in which the partial score sum for each interval is recorded. The sum of all the scores will then be the sum of this new column. Table 4.18 illustrates this procedure. The column labeled fX' is used to record the partial sums. This notation is based on the following usage: let X'_i be the coded class index for the ith row of the table and let f_i be the frequency for the ith row. Then, the product $f_i X'_i$ is equivalent to the sum of f_i values of X'_i. The sum of column fX' is

$$\sum_{i=1}^{k} f_i X'_i$$

where k is the number of rows, or score intervals, in the table. For grouped data, the formula for the coded arithmetic mean can be written

$$\mu = \frac{\sum_{i=1}^{k} f_i X'_i}{N}$$

For the example in Table 4.18 the column fX' sum is 38 and the coded arithmetic mean is $\mu' = 38/55 = .6909$. Transforming from μ' to μ yields $\mu = i\mu' - ic = 3(.6909) - 3(-10) = 2.0727 + 30 = 32.0727$. As a basis for comparison, the arithmetic mean can also be computed from the original scores since these are available in Table 4.1. From the original scores,

$$\sum_{i=1}^{55} X_i = 1{,}763 \quad \text{and} \quad \mu = \frac{1{,}763}{55} = 32.0653$$

Thus, for this example, the two results for the arithmetic mean differ very little. In other cases, this difference may be more substantial. When N is large and not too few score intervals are used in forming the frequency distribution, values of the arithmetic mean based on the original scores and on the frequency distribution can be expected to be almost identical.

The variance can be computed from a frequency distribution by making the same assumption concerning score values in an interval as was made for computing the arithmetic mean, and by using coded class indexes. The sum of the coded scores is the sum of the fX' column previously discussed. In order to find the sum of squared coded scores, we must compute the quantities $f_i X_i'^2$. This is done by multiplying each $f_i X_i'$ value by the corresponding X_i' value to form a fX'^2 column. The sum of this column is

$$\sum_{i=1}^{k} f_i X_i'^2$$

The formula for the variance of coded scores is then

$$\sigma^2 = \frac{\sum_{i=1}^{k} f_i X_i'^2 - \left(\sum_{i=1}^{k} f_i X_i' \right)^2 / N}{N}$$

To illustrate this procedure, Table 4.18 is reproduced along with the needed fX'^2 column (Table 4.19). The sums necessary to compute σ'^2 are

$$\sum_{i=1}^{k} f_i X_i' = 38 \quad \text{and} \quad \sum_{i=1}^{k} f_i X_i'^2 = 300$$

Then,

$$\sigma'^2 = \frac{300 - (38)^2/55}{55} = \frac{300 - 26.2545}{55} = \frac{273.7455}{55} = 4.9772$$

Transforming, we obtain $\sigma^2 = i^2 \sigma'^2 = 3^2(4.9772) = 9(4.9772) = 44.7948$. Also, $\sigma = \sqrt{\sigma^2} = \sqrt{44.7948} = 6.6929$. As a basis for comparison, the sum of scores and sum of squared scores from the raw data (Table 4.1) are

$$\sum_{i=1}^{55} X_i = 1{,}763 \quad \text{and} \quad \sum_{i=1}^{55} X_i^2 = 58{,}923$$

TABLE 4.19   *Use of coded class indexes to compute the variance*

Class index, X	X'	f	fX'	fX'²
39	3	14	42	126
36	2	11	22	44
33	1	9	9	9
30	0	7	0	0
27	−1	5	−5	5
24	−2	4	−8	16
21	−3	1	−3	9
18	−4	1	−4	16
15	−5	3	−15	75
		55	38	300

Then,

$$\sigma^2 = \frac{58{,}923 - (1{,}763)^2/55}{55} = \frac{2410.84}{55} = 43.8335 \quad \text{and}$$

$$\sigma = \sqrt{43.8335} = 6.6207$$

As was true for the arithmetic mean, the values derived from raw data and grouped data show close agreement.

The use of coded class indexes in computing the arithmetic mean and variance from a frequency distribution results in smaller values to work with and is a definite advantage when hand-computational methods are used.

Problems

4.1. Form a cumulative percentage distribution utilizing the following scores. Use a class interval of 3 with the lowest interval being 21–23; show the true limits, frequency, cumulative frequency, and cumulative percentage.

34	22	40	37	33	27	41	37
36	29	44	38	35	31	45	40
37	32	24	41	37	33	27	43
40	35	30	44	38	36	31	46
43	36	33	25	41	37	34	28
46	37	35	31	44	39	36	32
48	49						

4.2.   Construct a frequency polygon, a histogram, and an ogive based on the results of Prob. 4.1.

4.3.   Using the cumulative percentage distribution from Prob. 4.1, locate P_{60} and P_{80}.

4.4.   Using the *raw scores* presented in Prob. 4.1, find the mean, median, and mode of the distribution; find the same derived measures utilizing the grouped-data distribution.

4.5.   Using the raw scores presented in Prob. 4.1, find the range, semi-inter-quartile range, variance, and standard deviation of the distribution; find the same derived measures utilizing the grouped-data distribution.

4.6.   Using measures based on raw scores, calculate the skewness index for the distribution in Prob. 4.1; interpret the index with respect to its sign.

4.7.   A counseling group is composed of five boys whose heights, in inches, are 61, 64, 67, 70, 73.   What is the mean height of the group?   What are the variance and standard deviation of the heights?

4.8.   Mr. Jones went fishing and caught five fish; their lengths were 9, 15, 18, 21, and 27 inches.   What is the mean length?   What are the variance and standard deviation of the lengths?

4.9.   Which of the two groups of measures presented in Probs. 4.7 and 4.8 has the greater variability; upon what statistic do you base your response?

4.10.   Using the data from Prob. 4.7, perform a linear transformation by subtracting 50 from each score.   Compute the mean, variance, and standard deviation for the transformed scores.   How does this type of transformation affect the value of the derived measures?   State a generalization that describes this relationship.

4.11.   In order to reduce the magnitudes of a set of numbers, a student divided each original score by 5; using the transformed scores, he obtained a mean of 14 and a variance of 8.   What is the mean and standard deviation of the original scores?

4.12.   A student was asked to compute the mean and variance for some data. He arranged the data into intervals with a width of 5 units.   Since he had no calculator, he linearly transformed the class indexes by dividing by 5 and subtracting 10.   Computing with the transformed class indexes, he found the mean to be 4.986 and the variance to be 3.357.   What are the mean and variance for the grouped-data distribution (prior to transformation)?

Five

inferential methods for univariate measured data—one group

In this chapter we describe procedures for testing hypotheses when the data are measured. It is assumed that a single random sample has been drawn from a defined population and that the scores from the sample possess at least equal-interval properties. Basically, two types of null hypotheses are considered: (1) hypotheses concerning the arithmetic mean μ of the population; and (2) hypotheses concerning the variance σ^2 of the population.

Whereas null hypotheses concerning other characteristics of a population could be formulated and tested (e.g., hypotheses concerning skewness could be set up) it is generally sufficient to restrict attention to just these two types of null hypotheses. Focusing on the arithmetic mean and variance of a population for inferential purposes is justified on the basis of both practical and theoretical considerations. From a practical point of view, problems concerning populations are very often answerable in terms of the location of the population along a score axis. That is, we want to make statements concerning the average performance, attitude, etc., of a defined population. Similarly, the spread of scores in a population is often relevant to questions concerning the effects of instruction, changes in attitudes, etc. From the theoretical viewpoint, the arithmetic mean and variance play a central role in dealing with so-called "normal" distributions of scores. Normal distributions will be of major concern in this chapter, as we shall see.

In Sec. 5.1, the topic of normal distributions will be treated in detail and the rationale for their importance in applied statistics will be given. In Sec. 5.2, sampling distributions will be discussed and the rationale underlying the importance of normal distributions will be further developed. Secs. 5.3 to 5.5 treat specific inferential techniques related to the two major types of null hypotheses mentioned above.

5.1 Normal distribution theory

The so-called "normal distributions" or "normal curves" form a class of mathematical distributions that has properties fundamental to all advanced statistical theory. It should be pointed out that there is an infinite number of normal distributions and the phrase "the normal curve" is an incorrect usage, since it implies that there exists only one such distribution. In fact, there exists a normal curve for every possible combination of values for the arithmetic mean and variance of a distribution. Thus, there is a normal distribution with arithmetic mean of 12 and variance of 9; another with arithmetic mean of 22.603 and variance of 104.40; and so forth, for any of an infinite number of possibilities. However, all normal distributions have certain common properties that, in fact, define normal distributions. Mathematically speaking, a normal distribution is a function of the general type

$$Y = \left(\frac{1}{\sigma \sqrt{2\pi}}\right) e^{-.5[(X-\mu)^2/\sigma^2]}$$

where e is the transcendential number whose first six digits are 2.71828 and π is the ratio of the circumference to the diameter of a circle (that is, $\pi = 3.14159$ . . .). If μ and σ^2 are given specific numerical values, then a specific normal distribution is defined. For example, if μ is set at 0 and σ^2 at 1, we obtain the *unit normal distribution*

$$Y = \left(\frac{1}{\sqrt{2\pi}}\right) e^{-.5X^2}$$

Because of its simple form, the unit normal curve will be of interest to us. Any normal distribution can be graphed on ordinary rectangular coordinates by

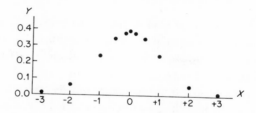

figure 5.1   Graph of selected X and Y points for
the unit normal distribution.

substituting values of X and solving for corresponding Y values.† Table 5.1 shows a few selected pairs of values for X and Y from the unit normal distribution. Figure 5.1 shows these same points plotted on a rectangular coordinate

TABLE 5.1   Selected X and Y values for the unit normal distribution

−3.0	.0044	.2	.3910
−2.0	.0540	.5	.3521
−1.0	.2420	1.0	.2420
−.5	.3521	2.0	.0540
−.2	.3910	3.0	.0044
0.0	.3989		

system.   If a smooth curve is drawn connecting these points (Fig. 5.2), we see what the distribution would look like if a much larger number of points were plotted.   Several characteristics of this curve should be noted.   (1) It is symmetric about the point $X = 0$; in general, any normal distribution is symmetric about the value of its arithmetic mean μ.   (2) Although the graph as shown extends only from -3 to $+3$, it actually extends infinitely in both directions; as we move further into the tails of the distribution, the curve gets closer and closer

† The value of X may be *any* positive or negative number.   Thus, the range of X is from $-\infty$ to $+\infty$.

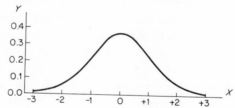

figure 5.2   Smoothed graph of the unit normal
distribution.

to the X axis but never touches it (that is, Y is never 0); this property is described by saying that the tails of the curve are asymptotic to the X axis.   (3) The curve is unimodal and its arithmetic mean, median, and mode have the same value (in this case, 0).   (4) If you study just the right half (or the left half) of the curve, you will note that the curve changes its direction of curvature as you progress from 0 to 3.   The exact point at which this change occurs is known as a *point of inflection;* in any normal curve, the points of inflection are at $\mu + \sigma$ and $\mu - \sigma$ (for the unit normal distribution, these are $0 + 1 = 1$ and $0 - 1 = -1$). (5) The total area under a normal distribution (i.e., between the curve and the X axis) is 1; this is true for any normal distribution, not just the unit normal distribution.   Also, since a normal distribution is symmetric, the half of the curve to the left of μ has area .5.

The normal distributions are important for applications in descriptive statistics because many empirical distributions show the same general shape and properties as do normal distributions.   For example, the distribution of intelligence scores for a large, heterogeneous population of elementary school children would be nearly symmetric and unimodal with inflection points in the right and left tails of the distribution.   Also, the proportions of cases in this population between various points, or above or below various points, will approximate very closely the corresponding areas under a normal distribution with the same arithmetic mean and variance.   However, the empirical distribution of intelligence scores will depart from normality in certain respects: (1) the distribution will have very definite maximum and minimum score values, so that the tails will not be asymptotic to the score axis; (2) only a finite number of scores will occur in the empirical distribution; hence, the perfectly smooth normal curve will not be obtained.

In any normal distribution, cumulative proportions and other relationships involving proportions are related in a relatively simple way to the arithmetic mean and standard deviation of the distribution.   For example, in a normal distribution, the 95th percentile always occurs at a distance of 1.645σ above the arithmetic mean; that is, $P_{95} = \mu + 1.645\sigma$.   Similarly, the 5th percentile occurs at $\mu - 1.645\sigma$ due to the symmetry of a normal distribution. Since these relationships will be very important in future applications of normal curve theory, a large number of them has been tabled (Tables F.1 and F.2). The use of these tables will now be explained in some detail.   In order to simplify the normal-curve tables, they are set up in terms of transformed scores. The transformation is of the form $z = (X - \mu)/\sigma$, where z is the transformed X score.   Note that the arithmetic mean of the z scores will be 0 and the variance and standard deviation will each be 1.[†]   In effect, applying a z-score

---

† The student should confirm this by applying the concepts of coding presented in the previous chapter.

transformation to every score in a normal distribution with arithmetic mean μ and standard deviation σ results in a distribution of z scores, which is normal with arithmetic mean of 0 and standard deviation of 1. That is, a z-score transformation of a normal distribution always results in a *unit normal distribution*.† Thus, Tables F.1 and F.2 are for the unit normal curve; however, we can always transform back from z scores to the original X scores by the relationship $X = \mu + \sigma z$. Tabling only the unit normal curve is for practical convenience, since information from these tables can always be referred back to any normal distribution.

Table F.1 gives the following information for each z score in the table. Column $A(Z)$ gives the proportion of cases in the distribution *between* z and 0. Column $B(Z)$ gives $.5000 - A(Z)$; that is, the proportion of cases in the distribution *above* z, for positive values of z, or *below* z, for negative values of z. Column $C(Z)$ gives the total proportion of cases *below* z, for positive values of z, or *above* z, for negative values of z [that is, $C(Z) = 1.0000 - B(Z)$]. Column $D(Z)$ gives the proportion of cases *between* symmetrically placed values of z [that is, $D(Z) = 2A(Z)$]. Column $E(Z)$ gives the proportion of cases *more extreme than* symmetrically placed values of z (that is, $[E(Z) = 1.0000 - D(Z)]$. And column $F(Z)$ gives the height of the ordinate at z (i.e., the value found by substituting z in the formula for the unit normal distribution). The figures in the column headings of Table F.1 illustrate the information about the unit normal curve by each column. Relationships among columns $A(Z)$, $B(Z)$, and $C(Z)$ are shown in Fig. 5.3 for a z score of $+1$ and in Fig. 5.4 for a z score of -1. Table F.2 is designed for finding various percentiles of the unit normal distribution. The column labeled P gives percentile ranks (in the form of proportions rather than percentages), column Z shows the corresponding z score, and column $F(Z)$ gives the height of the relevant ordinate. Thus, the 80th percentile can quickly be found to be a z score of .84.

† Of course, if the original distribution of X scores were not normal, transforming to z scores would result in a nonnormal distribution of z scores. The shape of the distribution of z scores is the same as the shape of the original distribution.

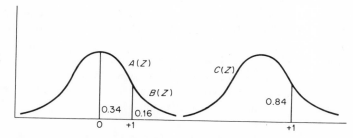

figure 5.3   *Area relationships under the unit normal curve, z = 1.*

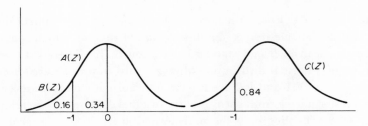

figure 5.4   Area relationships under the unit normal curve, z = −1.

Table F.1 can also be used to find the proportion of cases between any two z scores (not necessarily symmetrically placed). If both z scores have the same sign (i.e., both are positive or both are negative), and we call the larger one (in absolute value) z_1 and the smaller one (in absolute value) z_2, then the proportion of cases between them is† $C(z_1) - C(z_2)$. If the two z scores lie on opposing sides of 0, and we call one z_1 and the other z_2, then the proportion of cases between them is $A(z_1) + A(z_2)$.

For empirical score distributions that are *approximately* normal, the proportions in Tables F.1 and F.2 will be *approximately* accurate. Let us illustrate the use of these tables for the distribution of Stanford-Binet intelligence-scale scores (referred to as IQ scores). The general population arithmetic mean for such scores is 100 and the standard deviation is 16. The unit-normal-curve tables will be used to answer each of the questions below. Since a sketch of a normal curve with the information from the problem shown on the sketch is a useful device for solving problems involving normal distributions, this technique has been used in each of the examples.

1)   What proportion of the general population has an IQ of 130 or above? An IQ of 130 is a z score of $(130 - 100)/16 = 30/16 = 1.88$. Using column $B(z)$ of Table F.1, we find that the proportion of case above a z score of 1.88 (and, hence, above an IQ of 130) is .0301. That is, about 3 percent of the general population have IQs above 130 (Fig. 5.5).

2)   What proportion of the general population has IQs between 90 and 110? The z score for an IQ of 90 is $(90 - 100)/16 = -.62$ and for an IQ of 110 is $(110 - 100)/16 = .62$. Since these lie on opposite sides of 0, we let $z_1 = -.62$ and $z_2 = .62$. Then, $A(z_1) + A(z_2)$ is the required proportion. From Table F.1 we find $A(z_1) = .2324$ and $A(z_2) = .2324$. Thus, .4648, or about $46\frac{1}{2}$ percent of all IQs are between 90 and 110 (Fig. 5.6). Also, this result could be found immediately from column $D(Z)$ since z_1 and z_2 are symmetrically placed.

---

† If both z scores are negative, care must be exercised when following this rule; a z score of −1 is smaller in absolute value than a z score of −3.

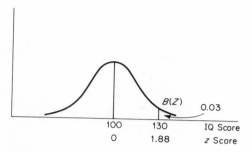

figure 5.5   Sketch for problem 1.

3)   What IQ score is the 90th percentile of the IQ distribution?   From Table F.2, the 90th percentile of the unit normal curve is 1.28.   The corresponding IQ score is $\mu + z\sigma$, or $100 + 1.28(16) = 120$ (Fig. 5.7).
4)   If the middle 75 percent of the general population is defined as having average IQ, between what IQ scores does this group fall?   Seventy-five percent of the population lies between $P_{12.5}$ and $P_{87.5}$.   From Table F.2, we find that $P_{12.5} = -1.1503$ and $P_{87.5} = 1.1503$.   Transforming these z scores to IQ scores yields $100 - 1.1503(16) = 82$ and $100 + 1.1503(16) = 118$.   Thus, the middle 75 percent of the IQ distribution falls between IQ scores of 82 and 118 (Fig. 5.8).

In order to simplify our discussion of proportions of cases under normal distributions, we introduce the following notation: $P(a < X < b)$ is the propor-

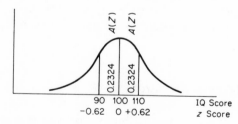

figure 5.6   Sketch for problem 2.

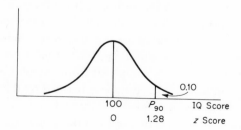

figure 5.7   Sketch for problem 3.

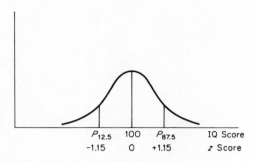

figure 5.8   Sketch for problem 4.

tion of X scores greater than a but less than b.   If we let X be an IQ score, then the result from problem 2 can be written $P(90 < X < 110) = .4648$.   Also, $P(X > a)$ and $P(X < b)$ are the proportions of cases above a and below b, respectively.   From problem 1, $P(X > 130) = .0301$; and from problem 3, $P(X < 120) = .90$.   It should be noted that all of the results from problems 1 through 4 are only approximate, since the actual distribution of IQ scores is not perfectly normal, although it is a reasonably close approximation.

One additional point must be considered before turning to a discussion of sampling distributions.   As we mentioned in Chap. 4, the area under a histogram or frequency polygon is related to the proportion of cases between selected scores along the score axis.   The same area relationships hold for normal curves. We will use the phrase "the area between z_1 and z_2" as synonymous with the phrase "the proportion of cases between z_1 and z_2."   Similarly, the area above z_1 is equivalent to the proportion of cases above z_1.

5.2   Sampling distributions

The concept of sampling distributions is basic to a thorough understanding of the mathematical and logical foundations underlying inferential statistics.   It is a difficult concept to master, but once it is mastered, the most advanced topics in applied statistics will become much more understandable.   For this reason, a detailed treatment of sampling distributions is presented in this section.

In general terms, a sampling distribution is the *theoretical* distribution of a specified derived measure computed from samples of fixed size.   A sampling distribution can be considered to be generated by the following process.   (1) A population of interest is defined and some derived measure is specified.   (2) A random sample of size n is drawn from this population and the derived measure is computed from the sample; a second random sample of size n is drawn and the derived measure is again computed from this sample. This process of random sampling and computation is repeated a large number of times (in theory, an infinite number of times). (3) The distribution of values for the derived measure is the sampling distribution of the derived measure. Fortunately, the properties of sampling distributions can often be determined by purely mathematical procedures, so repeated sampling is not necessary.

Let us illustrate this process for a population defined as the IQ scores of all students in a certain county. The derived measure of interest is, say, the arithmetic mean and we decide to draw random samples of size 10. To form the sampling distribution of the arithmetic mean, we must repeatedly draw random samples of size 10 from the defined population and compute the arithmetic mean of each sample. Of course, these sample arithmetic means will vary in value because of sampling variability. The distribution of these values of the sample arithmetic mean is the required sampling distribution of the arithmetic mean.

A sampling distribution can be generated for any derived measure and for any sample size. Thus, there is a *sampling distribution of the variance* for samples of size 20 drawn from a specified population; there is a *sampling distribution of the median* for samples of size 86 drawn from a specified population; and so forth. From the point of view of statistical inference, an important point is that for a wide variety of situations, the form and characteristics of a sampling distribution can be found by purely mathematical means without resort to the actual process of repeated sampling. Thus, for sample sizes that are not too small, the sampling distribution of the arithmetic mean is a normal distribution with arithmetic mean equal to the population arithmetic mean and variance equal to σ^2/n, where σ^2 is the population variance and n is the sample size. We shall return to consider the importance of this fact later.

A number of concepts related to the estimation of population values are based on sampling distributions. A value of a variable computed from a population is referred to as a *parameter;* a value of a variable computed from a random sample is referred to as a *statistic.* A basic problem in statistics is finding ways of using sample statistics to estimate population parameters. This may seem, at first sight, to be a simple problem. However, the obvious solution may or may not be the best choice for a specific parameter. Criteria for what constitutes a good estimator can be defined in terms of sampling distributions. Three of the most important criteria are unbiasedness, consistency, and efficiency:

1)   *Unbiasedness.* A sample statistic is described as an unbiased estimator of a population parameter if the arithmetic mean of the sampling distribution of the statistic is equal to the population parameter being estimated.

2)   *Consistency*. An estimator is described as consistent if as sample size is increased the variance of the sampling distribution of the statistic becomes progressively smaller. Thus, a consistent estimator "converges" to the value of population parameter for large sample sizes.

3)   *Efficiency*. This term is used when two or more estimators of the same population parameter are compared. An estimator whose sampling distribution has the smaller variance is described as more efficient than one with a larger variance.

These criteria can be applied to selecting estimators for two of the most important population parameters, the arithmetic mean and the variance. In the case of the arithmetic mean, a number of estimators are available and more than one of these is unbiased. For example, the sample arithmetic mean $\bar{X}$ is computed as

$$\bar{X} = \frac{\sum\limits_{i=1}^{n} X_i}{n}$$

This statistic is unbiased, since the arithmetic mean of the sampling distribution of $\bar{X}$ is the population arithmetic mean μ. It is also consistent. However, there are many other unbiased estimators of μ. For example, the sample median is an unbiased estimator of μ; also, a statistic defined as the midrange and computed as $(X_{max} + X_{min})/2$ is an unbiased estimator of μ. The major basis of choice among these available estimators of μ is in terms of efficiency. Figure 5.9 shows the forms of the sampling distributions for the sample arithmetic mean (*a*), sample median (*b*), and sample midrange (*c*). Each of these sampling distributions is centered about its arithmetic mean μ. However, the sample arithmetic mean $\bar{X}$ is the most efficient, since its sampling distribution shows the least variability. Thus, the sample arithmetic mean $\bar{X}$ is commonly

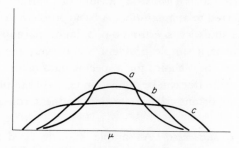

figure 5.9   Sampling distributions for (a) arithmetic mean, (b) median, (c) midrange.

used to estimate a population arithmetic mean μ. In this case, the obvious estimator turns out to be the best one available.

The problem of estimating the population variance σ^2 is different. The obvious estimator would be a sample statistic defined by analogy with σ^2; that is,

$$S'^2 = \frac{\sum_{i=1}^{n} (X_i - \bar{X})^2}{n}$$

However, S'^2 turns out to be a biased estimator; that is, the arithmetic mean of the sampling distribution of S'^2 is not *equal* to σ^2. Fortunately, this bias can be corrected quite simply. If we use $n - 1$ rather than n in the denominator of the sample variance formula, the resulting estimator is unbiased. Thus, an unbiased estimator of σ^2 is given by

$$S^2 = \frac{\sum_{i=1}^{n} (X_i - \bar{X})^2}{n - 1}$$

The quantity S^2 is always used in inferential statistics. We shall refer to S^2 as the "sample variance." Also, S^2 is a consistent estimator of σ^2 and it is the most efficient estimator.

As was mentioned earlier, the parameters of a sampling distribution can often be determined by purely mathematical means. By a useful mathematical theorem, known as the *central limit theorem*, it is known that, for "large" sample sizes, the sampling distribution of the arithmetic mean is a normal distribution with arithmetic mean of μ and variance of σ^2/n. This condition holds even when the population being sampled is *not* normal. In fact, the population can be of any shape as long as it has a finite variance and a finite mean. The central limit theorem is the basis for the importance of normal distributions in inferential statistics, since tests of inference can be derived from appropriate sampling distributions. Since we know that sampling distributions of the arithmetic mean are normal (for large n), all the properties of normal curves can be utilized without taking into consideration the form of the population being sampled. For this to hold in practice, the sample size must be considered large. This may be interpreted as (say) 20 or 30 cases for most situations.

To illustrate the interpretation of a sampling distribution of the arithmetic mean, we shall answer the following question. For the population defined as the IQ scores of all second-grade school children in New York City, what shape and parameters does the sampling distribution of the arithmetic mean have for samples of size 25 drawn from this population? Assume that the population arithmetic mean is 100 and the population variance is 16^2, or 256.

By the central limit theorem, this sampling distribution is normal, or nearly normal, since n is large. The arithmetic mean of the sampling distribution is the population mean, 100; the variance is $256/25 = 10.24$. Also, the standard deviation of the sampling distribution of the arithmetic mean is

$$\frac{\sigma}{\sqrt{n}} = \frac{16}{5} = 3.20$$

The standard deviation of the sampling distribution of the arithmetic mean is called the *standard error of the arithmetic mean* and is symbolized as $\sigma_{\bar{X}}$. Thus, $\sigma_{\bar{X}} = \sigma/\sqrt{n}$. By analogy, $\sigma_{\bar{X}}^2$ is used to symbolize the variance of the sampling distribution of the arithmetic mean. It should be noted that the variance of the sampling distribution of the arithmetic mean is generally smaller than the population variance (except for $n = 1$, when $\sigma_{\bar{X}}^2 = \sigma^2$). This is reasonable, since we would expect the arithmetic mean of n scores to show less variability than the individual scores themselves.

The sampling distribution of the variance can also be shown to be a normal distribution for sufficiently large sample sizes; also, the arithmetic mean of this sampling distribution is σ^2 if S^2 is used as the sample variance. However, the variance of the sampling distribution of the variance is generally not a simple expression. For cases where the population being sampled is itself normal, the variance of the sampling distribution of the variance is $(2/n)(\sigma^2)^2$.

In the remainder of this chapter, we treat specific inferential techniques for testing a hypothesis concerning either a population arithmetic mean or a population variance. The concept of the sampling distribution will recur often, so the student is advised to review the material presented in this section.

5.3   *Testing a hypothesis concerning* μ, *when* σ^2 *is known*

The use of sampling distributions to derive methods for testing hypotheses will now be illustrated. The situation described in this section rarely occurs in practice, but its simplicity makes a good starting point.

Assume that a random sample of scores obtained by measuring some variable has been drawn from a defined population. If these data are on an interval or ratio scale of measurement, we could compute the sample arithmetic mean $\bar{X}$ and the sample variance S^2. It may then be relevant to test hypotheses concerning parameters of the population from which the sample was drawn. For example, we could hypothesize that this population has some specified value for its arithmetic mean or for its variance. In this section, we consider the case in which the population variance is known to have a certain value;

but we wish to test the hypothesis that the population arithmetic mean has a specified value. In the next section (5.4), we consider the more general case in which the population variance is not known and the hypothesis of interest concerns the population arithmetic mean. In the final section (5.5), we consider the case of testing the hypothesis that the population variance has a specified value. For the first two cases, the null hypothesis is of the form H_0: $\mu = c$, where μ is the arithmetic mean of the population and c is some relevant numerical value. In the first case, σ^2 is known; in the second case, σ^2 is not known. For the third case, the null hypothesis is of the form H_0: $\sigma^2 = d$, where σ^2 is the variance of the population and d is some relevant numerical value.

Let us now turn to the first case. The general situation is as follows: we have drawn a random sample of n measures from a population of interest and have computed $\bar{X}$ from this sample.† The variance of this population is known to be some numerical value σ^2. The population arithmetic mean μ is unknown but hypothesized to have the numerical value c, that is, H_0: $\mu = c$. To test this hypothesis, we consider the nature of the distribution of possible sample arithmetic means from the population. This is the sampling distribution of the arithmetic mean. If n is not too small, this sampling distribution may be considered to be normal, according to the central limit theorem. Also, the arithmetic mean of the sampling distribution has the same value as the population arithmetic mean and the standard deviation of the sampling distribution is $\sigma/\sqrt{n}$. Thus, the sampling distribution is completely determined. Now, the observed sample arithmetic mean $\bar{X}$ can be located in the sampling distribution and the probability associated with its occurrence computed by use of the normal-curve tables. Figure 5.10 illustrates the general situation and includes an $\bar{X}$ value. The proportion of area that is shaded is the proportion of times a sam-

---

† We could, of course, also compute S^2 from the sample; but, as will be seen, we do not need this in the test of the hypothesis.

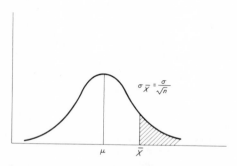

figure 5.10   *Sampling distribution of the arithmetic mean.*

ple arithmetic mean *as large as or larger than* $\bar{X}$ would occur by sampling variability alone. If this proportion could be calculated (which it can), we would know the *directional* probability associated with the null hypothesis. The probability is directional, since only values larger than $\bar{X}$ are included. The nondirectional situation will be deferred for the moment. To compute the proportion of cases in the shaded area of Fig. 5.10, it is only necessary to convert $\bar{X}$ to z-score form and use the $B(z)$ column of Table F.1. This will give us the proportion of area above z [that is, $P(a > z)$]. The sample arithmetic mean can be converted to z-score form by the procedures you have already learned; thus, $z = (\bar{X} - \mu)/(\sigma/\sqrt{n})$. Note that this formula has the same form as the z-score formula given earlier in this chapter; that is, we subtract the arithmetic mean of the sampling distribution from $\bar{X}$ and then divide by the standard deviation of the sampling distribution.

If z is so large that $P(a > z)$ is less than, say, .05, we can reject the null hypothesis and conclude that μ is greater than c.†

To summarize: for conducting directional tests, the null hypothesis is $H_0: \mu = c$. The alternative hypothesis of interest is *either $\mu > c$ or $\mu < c$* depending upon the specific problem being considered. In either case, we compute $z = (\bar{X} - \mu)/(\sigma/\sqrt{n})$. If our alternative hypothesis is $\mu > c$, we find $P(a > z)$ from Table F.1. If our alternative hypothesis is $\mu < c$, we find $P(a < z)$. If the probability is less than, say, .05, we can reject H_0 in favor of the alternative hypothesis; otherwise H_0 is accepted. The necessity of using Table F.1 can be avoided if the student remembers that 5 percent of the area of the unit normal curve is above 1.65 and 5 percent is below -1.65. If the alternative hypothesis is $\mu > c$ and if z exceeds 1.65, the H_0 can be rejected. Similarly, for a test at the 1 percent level of significance, the critical values are 2.33 and -2.33.

In the case of a nondirectional test, the alternative hypothesis is $\mu \neq c$ and we are interested in departures in either direction from μ. Thus, to compute the nondirectional probability associated with H_0, we compute the probability of an absolute *deviation* as large as or larger than $\bar{X} - \mu$. Since this can be in either direction (that is, $\bar{X} - \mu$ can be either positive or negative), we find the sum of $P(a > z)$ and $P(a < -z)$. If this sum is less than, say, .05, we reject H_0. Since a normal distribution is symmetric, $P(a > z) = P(a < -z)$. Hence, the nondirectional probability can be more conveniently computed as $2P(a > |z|)$. For reference, it should be learned that if $z = 1.96$, or $z = -1.96$,

---

† Since this is a directional-test situation, we would decide beforehand which direction to look for a difference. If we decided that only the situation in which μ was less than c was of interest, we would compute z and find $P(a < z)$. If $P(a < z)$ were less than, say, .05, we would conclude that μ was less than c.

then $P(a > |z|)$ is .025 and $2P(a > |z|) = .05$. These, then, are the critical values for a test at the 5 percent level of significance. Also, 2.58 and -2.58 are critical values for a nondirectional test at the 1 percent level of significance.

In practice, the one-sample z test is rarely applied, since the value of the population variance is usually not known. However, a contrived example will be used to illustrate the testing procedure. Assume that a researcher is studying the physical fitness of ninth-grade boys in a county-school system. He is especially interested in comparing these boys against national norm groups. For one measure of physical fitness, say, number of pushups, the national norm group shows an arithmetic mean of 7.2 and a standard deviation of 2.8. A random sample of 60 ninth-grade boys in the county-school system was tested and showed an arithmetic mean of 6.9 pushups. To decide if this is significantly different from the national level of performance, we could formulate the null hypothesis H_0: $\mu = 7.2$. That is, we hypothesize that the arithmetic mean of the total population of ninth-grade boys in the county is the same as the national normative value 7.2. The z test can be used because we know the general-population standard deviation is 2.8.† The standard error of the mean can be found to be $\sigma_{\bar{x}} = \sigma/\sqrt{n} = 2.8/\sqrt{60} = 2.8/7.75 = .36$. Then, the test statistic is $z = (\bar{X} - \mu)/\sigma_{\bar{x}} = (6.9 - 7.2)/.36 = -.30/.36 = -.83$. Since a value smaller than -1.96 would be needed to reject the null hypothesis at the nondirectional 5 percent level of significance, we would decide that there is no reason not to believe that our ninth-grade population of boys averages the same number of pushups as the national norm group.

5.4 *Testing an hypothesis concerning μ, when σ^2 is not known*

When the value of the population variance σ^2 is not known, it is no longer possible to compute the value of $\sigma_{\bar{x}}$, which is needed in the z test for the hypothesis H_0: $\mu = c$. The problem of constructing a valid test in this situation was of concern to mathematical statisticians around the turn of the century and the British statistician William Gossett, writing under the psuedonym "Student," derived a solution involving the so-called "t test" based on the t distributions. In order to utilize a t test, the sample standard deviation S is used

---

† Actually, the use of the standard deviation from the national norm group is not fully justified, since this is not the population of interest. The null hypothesis concerns the population comprised of all ninth-grade boys in the county and we do not know the standard deviation for this population. More properly, the test described in Sec. 5.4 is the appropriate one. We use the z test here merely to illustrate the computation and interpretation of a one-sample z test.

to estimate the unknown-population standard deviation σ. Then, the estimated standard error of the mean is computed: $S_{\bar{x}} = S/\sqrt{n}$. This value is used in place of $\sigma_{\bar{x}}$ in the test statistic. That is, $(\bar{X} - \mu)/S_{\bar{x}}$ is computed instead of $(\bar{X} - \mu)/\sigma_{\bar{x}}$. The ratio $(\bar{X} - \mu)/S_{\bar{x}}$ is called t, and probabilities associated with t values can be determined from available tables (Table F.3). However, there is an infinite number of theoretical t distributions, and the correct one must be chosen when performing the test of the hypothesis. As was true for the chi-square distributions, a t distribution exists for every different degree of freedom from 1 to infinity. For the one-sample t test of the hypothesis H_0: $\mu = c$, the correct degrees of freedom are $n - 1$, where n is the sample size. It should be noted that the t table is set up in the same manner as the chi-square table; that is, for each degrees of freedom, only certain t values are reported (for example, t values at the 5th, 95th, 99th, etc., percentiles). With the values from the t table, directional and nondirectional tests of hypotheses at the conventional 5 and 1 percent levels of significance can be made.

Two characteristics of the t table should be recognized. First, if you examine any one *column* of the t table, you will note that as the degrees of freedom increase, a smaller value of t is recorded in the column. This implies that for larger samples, smaller values of t will result in rejection of the null hypothesis. In a sense, the t distribution allows for the error involved in using $S_{\bar{x}}$ to estimate $\sigma_{\bar{x}}$; for larger samples, $S_{\bar{x}}$ is a better estimate of $\sigma_{\bar{x}}$ and less error of estimation is involved. Second, the last *row* of the table, for infinite degrees of freedom, is identical to corresponding values of the unit normal curve (that is, z values).

As was noted when we discussed the example at the end of the last section, the z test was not the best choice of technique. Since the standard deviation of the population of ninth-grade boys was really unknown, the sample standard deviation should be used to estimate σ and then the t test used to test the hypothesis H_0: $\mu = 7.2$. If the sample standard deviation were computed and found to be 2.4, the estimated standard error of mean would be $S_{\bar{x}} = S/\sqrt{n} = 2.4/\sqrt{60} = 2.4/7.75 = .31$. Then, $t = (\bar{X} - \mu)/S_{\bar{x}} = (6.9 - 7.2)/.31 = -.30/.31 = -.97$. Since $n = 60$, the degrees of freedom for this test are 59. Inspection of Table F.3 reveals that the null hypothesis must be accepted at the conventional levels of significance.

A common situation to which the one-sample t test has application involves the analysis of difference scores. Difference scores arise in educational research in a number of ways. One common example involves one group of individuals that is administered the same test on two different occasions (called the pretest and posttest). If some treatment has been applied to the individuals during the interval between pretest and posttest, it may be of interest to determine whether a gain, a loss, or no change has occurred as a result of the treatment. A t test based on the difference scores can be used to reach a decision. A second example involves matched pairs of individuals that are subjected to different treatments; difference scores for the pairs may be analyzed to determine whether or not the treatments result in different effects. An example of the first type is presented below.

A study involving the use of the pretest-posttest design can arise in many ways. Consider the case of a college English teacher at a large university who wants to evaluate the effectiveness of a new set of programmed instructional materials for teaching a course on elements of style to college freshman at this school. Over a period of years, a test of the students' ability to recognize good style has been used as a pretest and as a posttest in the regular course on writing style. In general, students have shown a gain between pretest and posttest as a result of the instruction. The arithmetic mean of this gain has been 8.72 score points over the years. The college teacher wants to decide if the gain would be different from this value using the new instructional materials. Since the cost of setting up the new programmed instructional procedures is fairly high, the teacher decides to test the materials on a sample of freshmen rather than on the entire class. For this purpose, he randomly selects 25 freshmen from the enrollees in the course on style and assigns them to programmed instruction in place of their regular classroom instruction. Each member of the sample takes the pretest, completes the programmed materials, and takes the posttest. Table 5.2 shows the pretest, posttest, and difference scores for the programmed instruction students.

TABLE 5.2   Pretest, posttest, and difference scores
for 25 students

Student number	Pretest	Posttest	Difference	Student number	Pretest	Posttest	Difference
1	55	63	8	13	44	55	11
2	46	64	18	14	44	57	13
3	58	65	7	15	45	60	15
4	49	68	19	16	24	22	−2
5	52	70	18	17	30	28	−2
6	36	41	5	18	21	34	13
7	34	45	11	19	31	39	8
8	22	48	26	20	25	40	15
9	45	53	8	21	69	74	5
10	45	51	6	22	64	75	11
11	43	54	11	23	53	80	27
12	40	57	17	24	70	82	12
				25	73	84	11
				Sums	1,118	1,409	291

We let X stand for a difference score, and the null hypothesis for this study can be written $H_0\colon \mu = 8.72$. The relevant population is all freshmen at the college (assuming that all freshmen must take the course on elements of style); and the null hypothesis states that if this entire population were instructed by the programmed materials, then the arithmetic mean of the difference scores would be 8.72 (i.e., the same as the previous gain under ordinary instruction). A nondirectional test is needed here, since outcomes either above or below 8.72 would be possible; and the teacher is interested in deciding whether the new instruction is better, worse, or no different from the regular instruction.

Using the data in Table 5.2, we find that the sum of the difference scores is 291 and their arithmetic mean is $291/25$, or 11.64.† The standard deviation of the difference scores is computed from the sum of the scores and the sum of the squared scores as follows:

$$S = \sqrt{\frac{4{,}575 - (291)^2/25}{24}} = \sqrt{\frac{4{,}575 - 3{,}387.24}{24}} = \sqrt{49.49} = 7.03$$

Then, the estimated standard error of the mean is

$$S_{\bar{x}} = \frac{S}{\sqrt{n}} = \frac{7.03}{\sqrt{25}} = \frac{7.03}{5} = 1.41$$

The t statistic can now be computed:

$$t = \frac{\bar{X} - \mu}{S_{\bar{x}}} = \frac{11.64 - 8.72}{1.41} = \frac{2.92}{1.41} = 2.07$$

Inspection of Table F.3 for 24 degrees of freedom reveals that $t = 2.07$ is just above the 97.5 percentile of the t distribution. Thus, with a risk of 5 percent, the college instructor would reject the null hypothesis and conclude that the programmed instruction does result in greater gains than conventional instruction.

5.5 Chi-square variance test

The statistical test considered in this section involves testing the hypothesis that a population variance has some specified value. That is, a random

---

† The arithmetic mean of the pretest scores is $1{,}118/25 = 44.72$ and of the posttest scores is $1{,}409/25 = 56.36$. The difference between these arithmetic means is $56.36 - 44.72$, or 11.64, which is the same as the arithmetic mean of the difference scores. It is quite easy to show that this must be true in general for difference scores. The student should prove this for himself. This fact provides a convenient check on the accuracy of the computation of the arithmetic mean of the difference scores.

sample of scores is available from a population of interest, and the problem is to decide whether the population variance has some specific numerical value. The null hypothesis is H_0: $\sigma^2 = d$.

The test of this hypothesis is very simple and merely involves computing the value of the sample variance S^2. Then, the ratio $(n - 1)S^2/\sigma^2$ is known to be distributed as a chi-square statistic with $n - 1$ degrees of freedom. It should be noted that in this application it is necessary to use both tails of the chi-square distribution to perform a nondirectional test. A chi-square distribution has its expected value, or arithmetic mean, at its degrees of freedom, which is $n - 1$ for the present test. Thus, if S^2 is less than σ^2, the value of $(n - 1)S^2/\sigma^2$ will be located in the left tail of the chi-square distribution. In order to reject the null hypothesis for values of S^2 that are significantly less than σ^2, a critical value in the left tail must be set up for nondirectional tests. Similarly, a critical value is set up in the right tail of the distribution in order to reject values of S^2 that are significantly larger than σ^2. Figure 5.11 shows these critical values for a nondirectional test at the 5 percent level of significance.

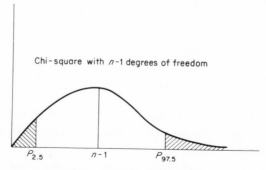

Chi-square with n-1 degrees of freedom

$P_{2.5}$ $n-1$ $P_{97.5}$

figure 5.11   *Critical values for a nondirectional test of the hypothesis H_0: $\sigma^2 = d$.*

Tests concerning variances are often of interest in educational research, especially in studies involving instructional methods. It may happen, for example, that some new instructional procedure does not raise the general level of achievement compared with previous techniques. Thus, a comparison of arithmetic means would reveal no significant differences. However, the new instruction might either significantly increase or decrease the variability of achievement among students. Knowledge of these kinds of effects may have implications for educational practice.

The computational mechanics of the one-sample variance test present no difficulties and will be illustrated for the difference-score data presented in Sec. 5.4 (Table 5.2). Assume that regular classroom instruction has yielded a variance of pretest-posttest difference scores of 144 (i.e., a standard deviation of 12). The relevant null hypothesis is, therefore, $H_0: \sigma^2 = 144$. The sample variance was 49.49 and the sample size was 25. Then,

$$\chi^2 = \frac{(n-1)S^2}{\sigma^2} = \frac{(25-1)(49.49)}{144} = \frac{1,187.76}{144} = 8.25$$

Choosing the conventional .05 level of significance, we see that Table F.4 shows that with 24 degrees of freedom, computed chi-square must be either larger than 39.36 or smaller than 12.40 to imply rejection of the null hypothesis (i.e., these are the 97.5 and 2.5 percentiles of the chi-square distribution with 24 degrees of freedom). Since 8.25 is less than 12.40, the null hypothesis is rejected, and we conclude that the programmed instructional procedures resulted in significantly *less* variability in difference scores than did conventional classroom instruction.

The techniques presented in this chapter are restricted to situations in which only one sample of subjects is utilized in a research study. It is much more usual for two or more different samples to be compared in one study; techniques for these situations are developed in Chap. 6.

Problems

5.1. For the unit normal distribution, find (a) $P(.75 < z < 1.68)$ and (b) $P(z > -.26)$.

5.2. Assuming a normal distribution of scores with a mean of 40 and a standard deviation of 10, (a) what proportion of scores are above 60 and (b) what proportion are between 30 and 50?

5.3. During manufacture, golf balls must pass several tests in order to be acceptable for marketing. One test measures the distance that a ball would have traveled after a specific impact. A manufacturing process produces balls that are normally distributed and travel a mean distance of 250 yards with a variance of 36. If a ball must travel between 235 and 260 yards to be acceptable, what proportion of balls from this process would meet the standard?

5.4. A random sample of size 20 is taken from a population with a hypothesized standard deviation of 15. The sample estimate of the variance is found to be 100. Can it be said with 90 percent confidence that the sample and population variances are consistent?

5.5. Give three of the most important criteria that constitute a good estimator for population parameters.

5.6. Assume that the population defined as the weights of all seventh-grade boys in the state of Maryland has a mean of 100 pounds and a variance of 400. What are the parameters of the sampling distribution of the mean for samples of size 25?

5.7. A fourth-grade classroom was given instruction on a topic in geography by means of programmed instruction.   If the class of 28 students shows a mean score of 20 on a standardized test that has a county-wide mean of 22 and standard deviation of 8.4, does the class mean differ significantly (at the .05 level) from the county mean score?

5.8.   Final-examination scores for a college statistics course have had a mean of 65 and a standard deviation of 7.5 for a number of years.   If a class of 25 students using a "new" method score a class mean of 69 and a standard deviation equal to that of all previous classes, can the instructor conclude that his present class has learned more?

5.9.   The following data represent pretest and posttest scores for a group of 5-year-old children who did not attend kindergarten.   The established mean gain on this test for children who do attend kindergarten is 3.0 units.   Test the hypothesis that this sample has a true population mean of 3.0 units and utilize the .01 level of significance.

Student number	Pretest score	Posttest score	Student number	Pretest score	Posttest score
1	13	16	10	4	14
2	13	13	11	3	14
3	13	15	12	3	15
4	10	13	13	3	13
5	8	15	14	3	15
6	6	17	15	2	13
7	6	15	16	2	13
8	5	13	17	1	14
9	5	13			

Six

methods for univariate measured data—two or more groups

Many important and interesting educational research problems involve comparing two or more groups of subjects in terms of their scores on some single measured variable. A "classical" example is that in which one group of subjects (the experimental group) is treated in a special way, while a second, control group, otherwise comparable to the experimental group, does not receive special treatment. If the two groups are compared on a relevant postmeasure, then the effect of the treatment can be determined. This experimental-vs.-control-group design can be easily generalized to cases involving additional groups of subjects. For example, a study might be aimed at determining the effect of distribution of study sessions upon achievement when programmed instruction is used as the teaching procedure. A unit of programmed instruction in mathematics is used, and the subjects are ninth-grade students. Randomly formed groups of subjects might be exposed to the following conditions:

GROUP 1   Studies the programmed instructional material for 2 hours at one session without a rest period.

GROUP 2 Receives two 1-hour study sessions, separated by a 10-minute break during which they engage in nonrelevant tasks.

GROUP 3 Receives four $\frac{1}{2}$-hour sessions, separated by 10-minute breaks as in group 2.

GROUP 4 Receives eight 15-minute sessions, separated by 10-minute breaks as in group 2.

If all groups take a posttest of achievement at the end of the last study session, then comparisons among the groups would give some evidence concerning the effects of distributing study sessions. These effects could involve differences in average performance (e.g., the arithmetic means of posttest scores might be compared) or they might result in more or less variability of scores among the groups (e.g., comparisons of variances of posttest scores could be made).

The use of several groups, rather than just two groups, is advantageous in many research settings. In general, multigroup designs permit efficient comparisons among different experimental treatments as well as comparisons of each with a control condition. Note that if the classical two-group, experimental-vs.-control-group design had been used in the above example, a total of six experiments would be required to make all of the possible comparisons two at a time.

In this chapter, techniques are presented for analyzing results from studies utilizing two or more groups of subjects. Only inferential procedures are considered because it is nearly always the case that sampling is used in practical situations of the types considered here. However, if population comparisons were appropriate in some research setting, the analysis would be especially simple. That is, compute the desired parameters for each population and compare them directly. Whatever differences occur are reliable, "significant" differences, and conclusions can be drawn immediately concerning the populations involved.

Tests for two general classes of hypotheses are considered. First, tests for equality of population variances are developed.† If the values of the same parameter in two or more populations are equal, then the populations are described as being *homogeneous* with respect to the parameter. Thus, the variance tests presented are often described as tests of *homogeneity of variance*. In symbols, the general form of the null hypothesis is $H_0: \sigma_1^2 = \sigma_2^2 = \cdots = \sigma_k^2$, where k groups are being compared. The second category of null hypotheses concerns population arithmetic means. The general form of the hypothesis

---

† Of course, if the variances of two or more populations are equal, then their standard deviations are also equal, since $\sigma = \sqrt{\sigma^2}$. The tests are based on variances rather than standard deviations, since the test statistic utilizes the variances.

of homogeneity for k population arithmetic means is $H_0\colon \mu_1 = \mu_2 = \cdots = \mu_k$. Tests for this hypothesis are performed by utilizing a procedure known as the *analysis of variance*. In this chapter, the variance tests are presented before the tests for arithmetic means. Why this order of presentation was chosen will become clear as the procedures for testing these classes of hypotheses are developed.

6.1 *Homogeneity of variance tests*

There are several available testing procedures for the hypothesis of homogeneity of variance. Of these, two are considered in this text. The first is appropriate for the case when only two samples are being compared; the second is more general and may be used for testing the homogeneity of three or more population variances.

For the two-sample homogeneity-of-variance test, the situation is one in which random samples are available from two populations. These samples may be of any size, two or larger. Of course, the larger the samples, the greater will be the sensitivity (or power) of the test to detect heterogeneous population variances. If the researcher were interested in deciding on the relationship between the magnitudes of the population variances, the relevant null hypothesis would be $H_0\colon \sigma_1^2 = \sigma_2^2$; and, in general, a nondirectional alternative, $\sigma_1^2 \neq \sigma_2^2$, would be appropriate.† It is interesting to note that the population variances need not be given a specific numerical value. The test depends upon the equality of σ_1^2 and σ_2^2 but not on their numerical values. The usual test statistic is very easy to compute; it is simply the *ratio* between the two variances computed from the samples, S_1^2/S_2^2. This ratio is known as an *F ratio;* the sampling distributions of F ratios have been worked out and tabled. These sampling distributions are not of a simple form, although they depend only upon the degrees of freedom for the two samples (i.e., upon $n_1 - 1$ and $n_2 - 1$). For every possible combination of sample sizes, there exists a unique F distribution with degrees of freedom $n_1 - 1$ and $n_2 - 1$. Table F.5 gives selected percentiles for certain of these F distributions. In order to use this table, it is necessary to note that it is organized in terms of numerator and denominator degrees of freedom; there are separate subtables for each numerator degrees of freedom, and within these subtables, we present the F values for different denominator degrees of freedom. The columns of the F table correspond to

---

† For notational purposes, subscripts will be used on population parameters and sample statistics for identification of groups. Thus, σ_1 is the standard deviation of population 1, S_2^2 is the variance of sample 2, n_1 is the size of sample 1, etc.

percentile ranks (presented as proportions rather than as percentages). Thus, if both the numerator and denominator have 5 degrees of freedom, we find that the 95th percentile of F is a value of 5.0503; similarly, the 2.5 percentile with numerator degrees of freedom of 8 and denominator degrees of freedom of 40 is .2604. Both directional and nondirectional tests can be carried out using these tables; thus, a significance test at the .05 level would utilize either the 95th or the 5th percentile values, as relevant, for a directional test and 2.5 and 97.5 percentile values for a nondirectional test.

Before considering the details of the F-ratio test for homogeneity of variance, some general characteristics of F distributions should be noted. The shape of an F distribution will show positive skew; a typical distribution is shown in Fig. 6.1. The right tail of the distribution is asymptotic to the hori-

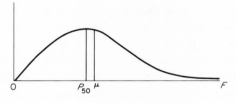

figure 6.1 *A representative F distribution.*

zontal axis, but the left tail stops at 0 (i.e., the origin). This is reasonable, since a ratio of positive numbers, such as S_1^2/S_2^2 must be, can never be negative in sign but can be 0 if $S_1^2 = 0$. Also, the arithmetic mean of an F distribution is always above its median, since the large values of F occur in the right tail of the distribution. An interesting relationship that is *not* obvious from the figure is that if you reverse the degrees of freedom for an F distribution, the new F distribution has percentiles which are *reciprocals* of the corresponding, symmetrically placed percentiles in the original distribution. In this context, P_α and $P_{1-\alpha}$ are considered to be symmetrically placed percentiles. Thus, for the F distribution with $n_1 - 1 = 20$ and $n_2 - 1 = 10$ degrees of freedom, $P_{95} = 2.7740$. The F distribution with $n_1 - 1 = 10$ and $n_2 - 1 = 20$ degrees of freedom will then have $P_5 = 1/2.7740$, or .3605. That is, the 5th percentile of the second distribution is the reciprocal of the 95th percentile of the first distribution.

The mechanics of carrying out an F-ratio test are straightforward. For each sample, compute the sample variance. Then, form the ratio $F = S_1^2/S_2^2$ and refer to the tabled F distribution with $n_1 - 1$ and $n_2 - 1$ degrees of freedom. For a nondirectional, .05-significance-level test, if F exceeds $P_{97.5}$ *or* if F is smaller than $P_{2.5}$, reject H_0; if F is between $P_{97.5}$ and $P_{2.5}$, accept H_0. If the null hypothesis is rejected, σ_1^2 and σ_2^2 can be ordered by inspecting S_1^2 and S_2^2.

If S_2^2 is larger than S_1^2, then $\sigma_2^2 > \sigma_1^2$ is implied. A practical example of this test will be given in Sec. 6.2. †

If samples from three or more populations have been randomly drawn and it is of interest to test $H_0: \sigma_1^2 = \sigma_2^2 = \cdots = \sigma_k^2$, then the F-ratio test is no longer applicable. Rather than devise a new test for the three or more group situation, it might seem proper to test homogeneity of variance by taking all possible *pairs* of samples and using the F-ratio test several times. Thus, for $k = 4$ samples, there are $4!/(2!2!)$ or six pairs of sample variances that could be used to form F ratios. Besides the mounting computational labor that accompanies this strategy as k becomes large (e.g., for $k = 8$, there are 28 pairs of variances to test), there is a more fundamental objection. Namely, the probability of a Type I error increases rapidly when such a *series* of statistical tests is performed. For just one F-ratio test, the Type I risk can be set at, say, .05. However, if two F-ratio tests are performed, then when H_0 is true, the probability of *at least one* false rejection in the two tests is $1 - .95^2$, or about .10. If three tests are run, each at the .05 level of significance, the *total* Type I error for all three tests is $1 - .95^3$, or about .14. In general, if a total of m tests, each at the αth significance level, is run, the total Type I error is $\alpha_m = 1 - (1 - \alpha)^m$. α_m is the risk of *at least one* false rejection in the set of m tests. Table 6.1 shows total Type-I-error rates for sets of one through ten tests for $\alpha = .05$ and $\alpha = .01$.

It is generally advisable to avoid the effects of compounding Type I errors that result from carrying out series of statistical tests such as those alluded to above. For the hypothesis $H_0: \sigma_1^2 = \sigma_2^2 = \cdots = \sigma_k^2$, there are procedures for testing the overall equality of the k variances by computing only one test statistic. Thus the Type-I-error risk can be set at .05 for rejecting the hypothesis of homogeneity of all k population variances. The test procedure presented in this textbook is known as Hartley's $F_{\max}$ test. Hartley's $F_{\max}$ test assumes that the k samples are of equal size. However, as we shall see, this test can be used under certain conditions for samples of unequal sizes.

To begin, let us assume that k samples, each of some constant size (for example, k samples each containing 10 observations) have been drawn randomly from k populations of interest. The null hypothesis is, of course,

$$H_0: \sigma_1^2 = \sigma_2^2 = \cdots = \sigma_k^2$$

---

† The F table in this textbook gives values from both tails of the F distributions. However, other sets of F tables that you use in the future might give only the right-tail values. If this happens, set up the F ratio with the *larger* variance in the numerator and the *smaller* variance in the denominator. In this way, the computed F value will always be larger than 1 and the right-tail values can be used. Be sure, however, to enter the table with the degrees of freedom in appropriate order and use the $1 - \alpha/2$ percentile value.

TABLE 6.1   *Total type-I-error rates for series of statistical tests*

Number of statistical tests, m	$\alpha = .05$; probability of at least one false rejection	$\alpha = .01$; probability of at least one false rejection
1	.05	.01
2	.10	.02
3	.14	.03
4	.19	.04
5	.23	.05
6	.26	.06
7	.30	.07
8	.34	.08
9	.37	.09
10	.40	.10

Compute the sample variances, $S_1^2, S_2^2, \ldots, S_k^2$, and compare them to locate the largest sample variance; call it S_{max}^2. Similarly, locate the smallest sample variance and call it S_{min}^2. The F_{max} statistic is simply $F_{max} = S_{max}^2/S_{min}^2$. The distributions of the F_{max} statistics are presented in Table F.6. To find 5 or 1 percent critical values associated with the null hypothesis, enter the segment corresponding to k, the number of groups or samples, and the row corresponding to degrees of freedom, $df = n - 1$, where n is the number of observations per sample. Thus, for 5 treatment groups, each comprised of 10 members, the appropriate k would be 5 and df would be $10 - 1$, or 9. Note that critical values for both .05- and .01-significance-level tests are presented in Table F.6. Also, only right-tail values of F_{max} are tabled. This is so because F_{max} must be 1 or larger and only an exceptionally large value of F_{max} implies rejection of the null hypothesis. A numerical example of this test is given later in this chapter.

If the samples are of different sizes, the F_{max} test may still provide a decision concerning H_0 *if* the following procedure yields consistent results. (1) Locate S_{max}^2, S_{min}^2 and compute F_{max} as above. (2) Compare the sample sizes, $n_1, n_2, \ldots, n_k$, and locate the largest sample size n_{max} and the smallest sample size n_{min} (n_{max} and n_{min} do *not* necessarily correspond to S_{max}^2 and S_{min}^2). (3) Enter Table F.6 for k and $df = n_{max} - 1$; if F_{max} is nonsignificant, then H_0 is accepted and the test is complete. If H_0 is rejected for $df = n_{max} - 1$, then (4) enter Table F.6 for k and $df = n_{min} - 1$; if H_0 is also rejected for

$df = n_{min} - 1$, conclude that H_0 is rejected and the test is complete. However, if H_0 is accepted for $df = n_{min} - 1$ but *rejected* for $df = n_{max} - 1$, the results are inconsistent and ambiguous. In this case, judgment concerning H_0 must be reserved and some other test applied.† Two examples will be used to illustrate this procedure. In the first, an unambiguous, consistent result is achieved; in the second, inconsistent results are found.

Example 1

For four random samples, the hypothesis $H_0: \sigma_1{}^2 = \sigma_2{}^2 = \sigma_3{}^2 = \sigma_4{}^2$ was set up; the sample sizes and variances were found to be as in the following table.

Sample	1	2	3	4
Size	10	12	8	16
Variance	21.67	18.96	12.31	20.63

S_{max}^2 is 21.67 and S_{min}^2 is 12.31. Then, $F_{max} = 21.67/12.31$, or 1.76. Since unequal sample sizes are involved, we locate $n_{max} = 16$ and $n_{min} = 8$. For $k = 4$ and $df = n_{max} - 1 = 15$, a value of F_{max} of *at least* 4.01 is required to reject the null hypothesis at the .05 level of significance. Since computed F_{max} is only 1.76, the null hypothesis of homogeneous variances is accepted. There is no need to check under $df = n_{min} - 1$, since *larger* critical values of F_{max} are required for smaller df values.

Example 2

For five random samples, the hypothesis is $H_0: \sigma_1{}^2 = \sigma_2{}^2 = \sigma_3{}^2 = \sigma_4{}^2 = \sigma_5{}^2$. The sample sizes and variances are:

Sample	1	2	3	4	5
Size	8	16	10	20	14
Variance	21.91	30.66	8.31	18.24	38.11

S_{max}^2 is 38.11 and S_{min}^2 is 8.31. Their ratio is $F_{max} = 38.11/8.31$, or 4.59. Then, $n_{max} = 20$ and $n_{min} = 8$. For $k = 5$ and $df = n_{max} - 1 = 19$, the 5 percent critical value is 3.71.‡ F_{max} exceeds this value; however, for $k = 5$ and $df = n_{min} - 1 = 7$, the 5 percent critical value is 9.70, and F_{max} is smaller than this value. Hence, the null is rejected for $df = n_{max} - 1$ but accepted for $df = n_{min} - 1$. These results are inconsistent, and the results of the test are ambiguous. A final decision would require the application of some other test, such as Bartlett's chi-square test, which does not assume equal sample sizes.

† A test that can be used when unequal sample sizes are present is Bartlett's chi-square test. Details of Bartlett's test can be found in Dayton [1970].

‡ Found by linear interpolation between degrees of freedom in the table.

6.2   The analysis of variance

From the outset, the student should avoid a confusion that arises from the name commonly used to describe the techniques discussed in this section. The "analysis of variance" is a set of procedures utilized in testing null hypotheses concerning population arithmetic means.   The name comes from the way in which the test is derived and does not refer to the hypothesis being tested. The terms "analysis of variance" and "homogeneity of variance" are sometimes confused by the beginning statistics student, so care should be taken to distinguish between these completely different procedures.

The importance of the analysis of variance in educational research can hardly be overemphasized.   From a frequency-of-usage viewpoint, the student will find the analysis of variance among the more popular statistical techniques encountered in research publications.   The widespread use of the analysis of variance springs from its great adaptability to a variety of research designs where hypotheses concerning population arithmetic means are being tested. In this book, we present only the simplest analysis of variance, the so-called one-dimensional analysis of variance.   In this design, k treatments of one type are each represented by one group of subjects.   The null hypothesis is

$$H_0: \mu_1 = \mu_2 = \cdot\ \cdot\ \cdot = \mu_k$$

That is, the arithmetic means of the populations from which the k samples were randomly drawn are equal to one another. †   In a more complex analysis of variance design, each group may receive two or more different *kinds* of treatments, and the effects of each different treatment dimension, as well as the interaction of treatment dimensions, can be assessed.   The variety of available analysis-of-variance techniques is very large, and a thorough *introduction* to this field requires more than a single course in statistics.   Needless to say, we can only scratch the surface in a beginning textbook.

In the beginning of this section, we present the basic assumptions of the one-dimensional analysis of variance and then algebraic derivations of some important statistics involved in the test of the null hypothesis.   Although the algebra used here is relatively simple, some students may find it difficult to appreciate fully the derivation.   If this is the case, a general understanding of the hypothesis test is nevertheless possible by studying the *logic* of the derivation and ignoring some of the finer details of the algebra.   It is hoped, however,

† When $k = 2$, the null hypothesis is $H_0: \mu_1 = \mu_2$.   The analysis of variance may be used to test this hypothesis.   Another procedure that may also be used to test this particular null hypothesis is the so-called "two-independent-samples t test."   The student will encounter references to this t test in his reading of research literature.   Details of this procedure are presented in Technical Appendix C.

that every student will make his best effort to follow the algebraic steps, since understanding of more advanced analysis-of-variance designs depends upon a thorough mastery of the one-dimensional design.

The conventional notation for scores from two or more groups of subjects is X_{ij}, where j identifies the group and i the position of the score in the group. Thus X_{35} is the third score in group 5. In general, n_j is the number of subjects in the jth group, and the total number of subjects in all k groups is

$$\sum_{j=1}^{k} n_j = n_t$$

This notation, and some extensions, are illustrated in Table 6.2. The sum of the scores in the jth group would be

$$\sum_{i=1}^{n_j} X_{ij}$$

As a shorthand, $X_{.j}$ is used in place of this sum. The arithmetic mean of the score in the jth group is

$$\frac{\sum_{i=1}^{n_j} X_{ij}}{n_j} \quad \text{or} \quad \frac{X_{.j}}{n_j}$$

TABLE 6.2   *Notation for one-dimensional analysis of variance*

	Group				
	1	*2*	*3*	$\cdots$	*k*
	X_{11}	X_{12}	X_{13}		X_{1k}
	X_{21}	X_{22}	X_{23}		X_{2k}
	X_{31}	X_{32}	X_{33}		X_{3k}
SUBJECTS	.	.	.		.
	.	.	.		.
	.	.	.		.
	$X_{n_1 1}$	$X_{n_2 2}$	$X_{n_3 3}$		$X_{n_k k}$
					Grand totals
Number of subjects	n_1	n_2	n_3	n_k	n_t
Sum of scores	$X_{.1}$	$X_{.2}$	$X_{.3}$	$X_{.k}$	$X_{..}$
Arithmetic mean	$\bar{X}_{.1}$	$\bar{X}_{.2}$	$\bar{X}_{.3}$	$\bar{X}_{.k}$	$\bar{X}_{..}$

and is symbolized $\bar{X}_{.j}$. The sum of all n_t scores in the table is $X_{..}$ and equals

$$\sum_{j=1}^{k} X_{.j}$$

Also, $\bar{X}_{..}$, the arithmetic mean of all the scores (also known as the *grand mean*), is $X_{..}/n_t$.

In the analysis of variance, it is assumed that each score is composed of the sum of three independent quantities. This assumption is written

$$X_{ij} = \mu + \tau_j + \epsilon_{ij}$$

where μ is the grand mean of the k populations; τ_j is the "effect" of being in the jth population (that is, $\tau_j = \mu_j - \mu$, where μ_j is the arithmetic mean of the jth population); and ϵ_{ij} is a random, normally distributed error term, unique to the ith individual in the jth group. The formula $X_{ij} = \mu + \tau_j + \epsilon_{ij}$ is known as the *structural model* for a score in the one-dimensional analysis of variance. The quantities in the formula can best be understood by studying a hypothetical population of scores. In Table 6.3, we present population

TABLE 6.3   *Hypothetical population scores*

	1	2	3	
	10	14	9	
	8	15	12	
	12	11	5	
		13	7	
		16		
				Grand totals
Number of subjects	3	5	4	12
Sum of scores	30	69	33	132
Arithmetic mean	10.00	13.80	8.25	11.00
	(μ_1)	(μ_2)	(μ_3)	(μ)

values for three small populations. Using population symbols, the arithmetic mean of the first population is $\mu_1 = 30/3 = 10.00$, and so forth. The grand mean μ is 132/12, or 11.00. Consider the score $X_{52} = 16$, the fifth score in the second group. From the structural model, $X_{52} = \mu + \tau_2 + \epsilon_{52}$. The grand mean μ is 11.00; the effect of group 2, τ_2, is $13.80 - 11.00 = 2.80$. Since $X_{52} = 16$, we can write $16 = 11.00 + 2.80 + \epsilon_{52}$. Hence, ϵ_{52} must be $16 -$

11.00 − 2.80, or 2.20 (which is equivalent to $X_{ij} - \mu_j$, or $16 - 13.80 = 2.20$). Similarly, the structural model can be interpreted for any other score in the table. For score X_{33}, the student should confirm that $\mu = 11.00$, $\tau_3 = -2.75$, and $\epsilon_{33} = -1.25$.

Of course, when random samples are involved, the terms μ, μ_j, and ϵ_{ij} cannot be computed directly. However, they can be estimated by the corresponding sample quantities. Thus, $\bar{X}_{..}$ *estimates* μ; $\bar{X}_{.j} - \bar{X}_{..}$ *estimates* τ_j, and $X_{ij} - \bar{X}_{.j}$ estimates ϵ_{ij}.

The null hypothesis for the one-dimensional analysis of variance is $H_0: \mu_1 = \mu_2 = \cdots = \mu_k$. If the null hypothesis is true, this implies that each τ_j is 0. This is obvious, since in this case the grand mean μ is the same as μ_1, μ_2, etc. Thus, the null hypothesis could also be written

$$H_0: \tau_1 = \tau_2 = \cdots = \tau_k = 0$$

This can be interpreted as meaning that a given population has no effect on the value of a score in it, beyond the general grand mean effect μ. In other words, a score is no different, on the average, because it is from any specific population.

In order to test the null hypothesis, some additional assumptions concerning the error terms ϵ_{ij} must be made. It is assumed that the ϵ_{ij} values in any one group are normally and randomly distributed, with arithmetic mean of 0 and with constant variance from group to group.† The constant variance assumption is equivalent to assuming homogeneity of variance for the k populations. This is true because both μ and τ_j are fixed quantities within any one group. Hence, the only source of variability within a group is the different values of ϵ_{ij} for different members of the group. If $\sigma_1^2 = \sigma_2^2 = \cdots = \sigma_k^2$, this implies that the errors ϵ_{ij} are independent of the treatment effects τ_j. That is, populations with large effects have errors no larger than populations with small effects, and vice versa. This condition is often expressed by saying that the errors are uncorrelated with the treatments. Since the analysis of variance assumes homogeneity of variance, this assumption can be tested by applying a homogeneity-of-variance test before applying the analysis of variance. That is, the null hypothesis $H_0: \sigma_1^2 = \sigma_2^2 = \cdots = \sigma_k^2$ is set up and tested by the procedures of Sec. 6.1. This will be discussed in more detail after the analysis of variance is more fully developed.

Since the population variances are assumed to be equal to one another, the symbol σ^2 will be used to refer to the common value of these population

† In order to derive the test used in the analysis of variance, normality of distribution and homogeneity of variance must be assumed. When analyzing real data, as will be pointed out later, these assumptions can sometimes be violated without invalidating the usefulness of the test.

variances. A test of the null hypothesis $H_0: \mu_1 = \mu_2 = \cdots = \mu_k$ is derived by finding two different estimates of the common-population variance σ^2. One of these estimates is always an unbiased estimate of σ^2. The other is an unbiased estimate of σ^2 *only* if the null hypothesis concerning the population arithmetic means is true. If the μ_j values differ from one another, then this second variance tends to be an overestimate of σ^2. Thus, by comparing these two estimates, the reasonableness of the null hypothesis can be assessed. These two different estimates of σ^2 are obtained by a process of dividing the total variability of the scores into two separate components. For algebraic convenience, we begin by showing how the total sum of squares† is separated into two components and then converted to variances. Also, the derivation that follows assumes that the samples are of the same size; that is,

$$n_1 = n_2 = \cdots = n_k$$

The symbol n will be used to represent this constant-sample size. The results of the derivation also hold for groups of unequal size, and the formulas presented for the test of the null hypothesis may be used for samples of equal or unequal sizes.

If we ignore the treatment groups completely, we have a set of nk scores. The sum of squares for this total group of scores is

$$\sum_{j=1}^{k} \sum_{i=1}^{n} (X_{ij} - \bar{X}_{..})^2$$

where the double-summation operator is necessary to sum over both columns and rows of the data table. This quantity is known as the *total sum of squares* and is denoted SS_T. We proceed, now, to divide up, or "partition," this total sum of squares into additive components. We begin by adding and subtracting $\bar{X}_{.j}$ inside the parentheses:

$$SS_T = \sum_{j=1}^{k} \sum_{i=1}^{n} (X_{ij} - \bar{X}_{.j} + \bar{X}_{.j} - \bar{X}_{..})^2$$

Of course, adding and subtracting a constant results in a net gain of 0 and has no effect on the equation. Then, the terms can be grouped and the squaring operation carried out:

$$SS_T = \sum_{j=1}^{k} \sum_{i=1}^{n} [(X_{ij} - \bar{X}_{.j}) + (\bar{X}_{.j} - \bar{X}_{..})]^2 = \sum_{j=1}^{k} \sum_{i=1}^{n} [(X_{ij} - \bar{X}_{.j})^2$$
$$+ 2\,(X_{ij} - \bar{X}_{.j})(\bar{X}_{.j} - \bar{X}_{..}) + (\bar{X}_{.j} - \bar{X}_{..})^2]$$

† You should recall that "sums of squares" is the technical term for the sum of squared deviations from the arithmetic mean.

Applying the summation operators to the separate terms results in

$$SS_T = \sum_{i=1}^{k} \sum_{i=1}^{n} (X_{ij} - \bar{X}_{.j})^2 + 2 \sum_{j=1}^{k} \sum_{i=1}^{n} (X_{ij} - \bar{X}_{.j})(\bar{X}_{.j} - \bar{X}_{..})$$

$$+ n \sum_{j=1}^{k} (\bar{X}_{.j} - \bar{X}_{..})^2$$

The last term on the right is constant with respect to the i subscript [that is, $(\bar{X}_{.j} - \bar{X}_{..})$ does not contain i as a subscript anywhere in it]; hence, it is equivalent to summing a constant n times and the result

$$n \sum_{j=1}^{k} (\bar{X}_{.j} - \bar{X}_{..})^2$$

is obtained.

The middle term on the right can be shown to always equal 0 and can hence be dropped from the equation. This is so because the term $(\bar{X}_{.j} - \bar{X}_{..})$ is constant with respect to the i subscript. Hence, the middle term can be written as

$$2 \sum_{j=1}^{k} (\bar{X}_{.j} - \bar{X}_{..}) \sum_{i=1}^{n} (X_{ij} - \bar{X}_{.j})$$

However, each of the sums

$$\sum_{i=1}^{n} (X_{ij} - \bar{X}_{.j})$$

is equal to 0 because the sum of the deviations of the scores within one group about the arithmetic mean of that group is always 0.

Dropping the middle term, we obtain

$$SS_T = \sum_{j=1}^{k} \sum_{i=1}^{n} (X_{ij} - \bar{X}_{.j})^2 + n \sum_{j=1}^{k} (\bar{X}_{.j} - \bar{X}_{..})^2$$

These two terms each have relatively simple interpretations. The first term can be rewritten for emphasis as

$$\sum_{j=1}^{k} \left[\sum_{i=1}^{n} (X_{ij} - \bar{X}_{.j})^2 \right]$$

When $j = 1$, the term in the brackets is the sum of squares for group 1; when $j = 2$, it is the sum of squares for group 2; and so forth. Hence, the entire term is the sum of the separate sums of squares for the k samples. In effect, it is a "pooled" sum of squares, and is known as the *error sum of squares* (SS_{error}). The name is appropriate because each separate sum of squares for one of the samples can be converted to a variance by dividing by $n - 1$; this variance is then, an unbiased estimate of the corresponding population variance. However, since we assumed homogeneity of variance, the separate sums of squares can be pooled to obtain a single combined estimate of the common-population variance σ^2. Since each group has $n - 1$ degrees of freedom, the total degrees of freedom over k groups is $n_t - k$.† Thus, SS_{error} can be converted to an estimate of the population variance σ^2 by dividing by $n_t - k$. This estimate of σ^2 is unbiased whether or not the null hypothesis is true. It assumes only that the population variances are homogeneous and makes no assumptions concerning the population arithmetic means. In analysis-of-variance terminology, the result of dividing a sum of squares by its degrees of freedom is known as a *mean square;* a mean square is always a variance and the two terms may be considered as being synonymous. Thus, $SS_{error}/(n_t - k)$ is the error mean square and is written MS_{error}.

The second term resulting from the partitioning of the total sum of squares is

$$ n \sum_{j=1}^{k} (\bar{X}_{.j} - \bar{X}_{..})^2 $$

and is called the treatment sum of squares, SS_{treat}. This term can also be interpreted, but the reasoning is a little more involved. *If* the null hypothesis is true, then $\mu_1 = \mu_2 = \cdots = \mu_k$. Also, $\bar{X}_{.1}$ is an estimate of μ_1; $\bar{X}_{.2}$ is an estimate of μ_2; and so forth. However, $\mu_1 = \mu_2 = \cdots = \mu_k$ implies that each $\bar{X}_{.j}$ is estimating the same population value μ. Thus, the sample arithmetic means, $\bar{X}_{.1}, \bar{X}_{.2}, \ldots, \bar{X}_{.k}$, represent k separate estimates of the same population arithmetic mean value. Now, consider the following problem: how could you estimate the variance of the sampling distribution of arithmetic means based on samples of size n from a population with arithmetic mean μ and variance σ^2? The procedure you learned previously was to draw one random sample and compute S^2/n as the estimate of $\sigma_{\bar{X}}^2$. Another approach would be to draw several random samples and compute the sample arithmetic mean of each. Then, the variance of these sample arithmetic means would also estimate $\sigma_{\bar{X}}^2$. In fact, the expression

$$ \sum_{j=1}^{k} (\bar{X}_{.j} - \bar{X}_{..})^2 $$

is equivalent to the numerator of a variance of this type; $\bar{X}_{..}$ is the arithmetic mean of the $\bar{X}_{.j}$ values. Each $(\bar{X}_{.j} - \bar{X}_{..})$ is a deviation from the arithmetic mean, and

$$ \sum_{j=1}^{k} (\bar{X}_{.j} - \bar{X}_{..})^2 $$

† For groups of equal size, this is the same as $nk - k$, or $k(n - 1)$.

is a sum of squares. Since there are k samples, the degrees of freedom are $k - 1$ and

$$\frac{\sum_{j=1}^{k} (\bar{X}_{.j} - \bar{X}_{..})^2}{k - 1}$$

estimates $\sigma_{\bar{X}}^2$. However, the original term, SS_{treat}, is n times this sum of squares. But $\sigma_{\bar{X}}^2 = \sigma^2/n$. Thus,

$$SS_{\text{treat}}/(k - 1) = \frac{n \sum_{j=1}^{k} (\bar{X}_{.j} - \bar{X}_{..})^2}{k - 1}$$

will estimate σ^2 rather than $\sigma_{\bar{X}}^2$. This second estimate of σ^2 is the treatment mean square, MS_{treat}. It is an unbiased estimate of σ^2 *only if* the null hypothesis is true, since we assumed each sample arithmetic mean estimates a common-population arithmetic mean.

In summary, we found that the MS_{error} is an unbiased estimate of σ^2 whether or not the null hypothesis is true; the MS_{treat}, on the other hand, is an unbiased estimate of σ^2 only if the null hypothesis is true. If H_0 is not true, the MS_{treat} will tend to be larger than σ^2, since sample arithmetic means from populations with different arithmetic means will vary more than those from populations with a common arithmetic mean. This becomes clear if you consider an extreme case, such as three populations with equal arithmetic means versus three with arithmetic means of 100, 200, and 300, respectively.

From Sec. 6.1, you will recall that a ratio of two variances is an F distribution. Hence, if we compute $MS_{\text{treat}}/MS_{\text{error}}$, this is an F statistic with $k - 1$ and $n_t - k$ degrees of freedom. If the null hypothesis is true, this ratio tends to 1 in the long run; if the null hypothesis is untrue, this ratio will tend to be larger than 1. Therefore, a directional F test may be used to test

$$H_0: \mu_1 = \mu_2 = \cdots = \mu_k$$

Note that values of F less than 1 never imply rejection of H_0. When F is less than 1, this must be attributed to sampling variability. When F is large, however, this casts doubt on the null hypothesis. Critical values for F can be found in Table F.5.

Computational formulas for the analysis of variance

Computation of the sums of squares for the one-dimensional analysis of variance is greatly simplified by using special computational formulas in place of those given above. Derivation of these formulas is left as an exercise for the interested student. The total sum of squares SS_T can be found as

$$SS_T = \sum_{j=1}^{k} \sum_{i=1}^{n_j} X_{ij}^2 - \frac{X_{..}^2}{n_t}$$

The first term is the sum of the squared scores over the entire data table; $X_{..}^2$ is the square of the grand sum, and n_t is the total number of observations in the table.

The treatment sum of squares SS_{treat} can be found as

$$SS_{\text{treat}} = \sum_{j=1}^{k} \left(\frac{X_{.j}^2}{n_j} \right) - \frac{X_{..}^2}{n_t}$$

The first term is computed by squaring each column sum, dividing the result by the number of scores in the column, and summing over all columns.

The error sum of squares is most conveniently found by subtraction: $SS_{\text{error}} = SS_T - SS_{\text{treat}}$.

These results, as well as the conversion to mean squares and the computation of the F ratio, are usually summarized in a so-called "analysis-of-variance table." The general form of an analysis-of-variance table is shown in Table 6.4.

TABLE 6.4   *Analysis-of-variance summary table*

Source	Degrees of freedom	Sum of squares	Mean square	F
Treatment	$k - 1$	SS_{treat}	MS_{treat}	$MS_{\text{treat}}/MS_{\text{error}}$
Error	$n_t - k$	SS_{error}	MS_{error}	
Total	$n_t - 1$	SS_T		

The treatment and error degrees of freedom should add up to the total degrees of freedom. If they do not, a mechanical error in arithmetic should be searched for. However, the fact that the SS_{treat} and SS_{error} add up to the SS_T should *not* be taken as evidence of the correctness of analysis-of-variance computations. Remember that the SS_{error} was obtained by subtraction; and if either the SS_T or the SS_{treat} contain computational errors, these can only be found by rechecking the original computations. Since errors can easily creep into any long series of computations, it is strongly advocated that all analysis-of-variance computations be checked thoroughly. In general, the use of a calculating machine will greatly reduce the likelihood of error; and, in any case, the added speed of such a device makes the rechecking procedure less wearisome than when hand-computational procedures are followed.

Since the analysis of variance assumes that the samples come from populations with equal variances, this assumption of homogeneity of variance should be tested prior to carrying out the analysis of variance. For $k = 2$, the F-ratio test should be applied to the sample variances; for three or more samples, Hartley's F_{max} test may be used. If the hypothesis of homogeneity of variance if accepted, then the analysis of variance may be applied with assurance that the equal-variance assumption is satisfied. If the hypothesis of homogeneity of variance is rejected, however, several possibilities must be considered. First, if the samples are of equal, or nearly equal, size, there is strong empirical evidence that violating the homogeneity-of-variance assumption has only a mild effect on the analysis of variance.† Specifically, the tabular F values do not give tests that are exactly at the .05 or .01 levels of significance, but the risk of a Type I error will not greatly differ from these levels (e.g., between .04 and .06 when .05 tabular critical values are used). Thus, if the samples are of similar sizes, the analysis of variance may be used whether or not the variances are homogeneous. This is an important liberalization of the analysis of variance and should be considered during the design of experiments. If possible, always set up groups of equal size; then, even if heterogeneous variances occur, the analysis of variance may be properly applied to the data. A second consideration arises if the samples differ considerably in size and heterogeneous variances are found. It may be possible to find a transformation of the scores such that the variances of the transformed scores will show homogeneous variances. Commonly used transformations involve square roots, reciprocals, logarithms, or trigonometric functions of the original scores. As far as possible, the necessity for transforming should be anticipated in the design of an experiment and the appropriate transformation preplanned. The details of selecting an appropriate transformation are too complex for treatment in an elementary textbook, but the student should be aware that this general approach may be used to overcome the problem of heterogeneity of variance when it occurs. A third consideration when heterogeneous variances are present is the possibility of converting the data to ranks and using a test that assumes only ordinal data. This sidesteps the variance problem, but there are certain disadvantages to such a strategy. The null hypothesis for the ordinal-data test will be different from that for the analysis of variance. If the research design calls for a comparison of arithmetic means, the substitution of a different null hypothesis may not provide the most appropriate answer to the research prob-

† Reports of empirical evidence bearing on the robustness of the analysis of variance to unequal variances may be found in Boneau [1960] and Lindquist [1953].

lem being investigated. Also, if the data are equal-interval to begin with, it seems unfortunate to reduce them to a lower-level scale of measurement before analysis. In effect, some information contained in the equal-interval data is discarded during the conversion to ranks. However, conversion to ranks may be necessary if a transformation is unsuccessful in establishing equality of variances; it is better to use an ordinal-data test than to violate flagrantly the assumptions of an equal-interval-data test. Moreover, some of the ordinal-data tests are nearly as powerful for their null hypotheses as are the corresponding equal-interval-data tests.

Before moving on to an example of the computations and interpretations involved in the one-dimensional analysis of variance, one additional problem needs to be discussed. If the null hypothesis is accepted, the interpretation of the analysis of variance is straightforward; that is, we decide that the samples come from populations with equal arithmetic means and that the effects of the different treatments did not result in significant, reliable differences among the sample arithmetic means. However, if the null hypothesis is rejected, the situation is much less clear-cut. When we decide that the null hypothesis is untenable, there is usually a variety of ways in which this can be interpreted. For example, if the null hypothesis is $H_0: \mu_1 = \mu_2 = \mu_3$ and if this hypothesis is rejected on the basis of the analysis of variance, then several possibilities exist with respect to the population arithmetic means. These may be summarized by the nondirectional alternatives

$$(\mu_1 = \mu_2) \neq \mu_3 \qquad (\mu_1 = \mu_3) \neq \mu_2$$
$$(\mu_2 = \mu_3) \neq \mu_1 \qquad \mu_1 \neq \mu_2 \neq \mu_3$$

In the first case, populations 1 and 2 have equal arithmetic means but population 3 is different from either of these; similarly, population 2 or 1 may be the one with the divergent arithmetic mean. Also, all three arithmetic means may have different values, as in the fourth case. When the null hypothesis is rejected, it is still necessary to carry out additional steps to decide specifically where the differences between population arithmetic means are located. Of course, when the sample sizes are equal we can decide immediately that the sample with the largest arithmetic mean and the sample with the smallest arithmetic mean do differ significantly and, hence, that the corresponding population arithmetic means do differ. But beyond this, any remaining differences are not obvious and must be tested for in some systematic manner. A consideration of specific procedures for this purpose is beyond the scope of an introductory textbook of this type; once again, our intent in raising this issue was to alert the student to its existence and, perhaps, also to provide some motivation for his continued study of statistical methodology.

Computation of the analysis of variance

The data in Table 6.5 are from a small experiment that was designed to study the effects of different kinds of feedback information upon performance of a memory-learning task involving meaningful verbal material. Forty subjects were recruited from college English classes at one university and ten were

TABLE 6.5   *Final-test scores for four groups of subjects in a learning experiment*

	Treatment group					
	1	2	3	4		
	3	1	2	6		
	6	5	4	2		
	2	0	8	5		
	0	2	5	7		
	1	0	4	1		
	4	4	6	2		
	2	3	3	3		
	5	3	6	6		
	1	1		5		
		2		3		
					Grand totals	
Number of subjects	9	10	8	10	37	(n_i)
Sum of scores	24	21	38	40	123	$(X..)$
Sum of squared scores	96	69	206	198	569	$(\Sigma\Sigma X_{ij}^2)$
Arithmetic mean	2.67	2.10	4.75	4.00	3.32	$(\bar{X}..)$
Variance	4.00	2.77	3.64	4.22		

randomly assigned to each of the four treatment conditions. However, owing to a number of reasons (e.g., illness), not all subjects completed the experimental tasks and, hence, the final number in some groups was less than 10. The treatments assigned to the groups are described below:

GROUP 1   Studied a long poem (selected because it was unlikely to be familiar to college freshmen) for 10 minutes. A line of the poem was exposed for 10 seconds to the subjects by a special mechanical device.

The subject then recited the line as well as he could and then another line was exposed. This process was repeated for all 30 lines of the poem. No feedback information related to correctness of the students' recitals was given.

GROUP 2 Same procedure as group 1 except that at the end of the study session, the student was told how many lines he had recited correctly.

GROUP 3 Same procedure as group 1 except that the subject was told whether his line recital was correct immediately after each recital.

GROUP 4 Same procedure as group 1 except that the subject was corrected by interjected comments during his line recital whenever he made errors. If his recital was correct, he was told this.

At the end of their study session, each group took a test that asked them to write the poem, line by line, from memory. The score for a subject was simply the number of completely correct lines he was able to write, regardless of their order. The numbers in Table 6.5 are the scores for the four groups of subjects. The column sums, sums of squared scores, arithmetic means, and variances are shown at the foot of the table. The grand sums and the grand mean are shown at the right of the data table.

The first step in the analysis is to test for homogeneity of variance. In the present case, this step is not critical because even if homogeneity of variance were rejected, the analysis of variance could still be applied to groups of nearly equal sizes. It is of interest, however, to test for homogeneity of variance in order to decide if the four procedures resulted in significantly different amounts of variability in learning. The null hypothesis is $H_0: \sigma_1^2 = \sigma_2^2 = \sigma_3^2 = \sigma_4^2$. Comparing the sample variances, we find that $S_{max}^2 = 4.22$ and $S_{min}^2 = 2.77$. Then, $F_{max} = S_{max}^2/S_{min}^2 = 4.22/2.77 = 1.52$. Since unequal sample sizes exist, we find $n_{max} = 10$ and $n_{min} = 8$. Entering Table F.6 with $k = 4$ and $df = n_{max} - 1 = 9$, we find that an F_{max} value of at least 6.31 must occur for rejection of the null hypothesis at the .05 level of significance. Since the computed F_{max} is less than the tabular critical value, we accept the equality of the population variances and decide that the four procedures do not have differential effects on the spread of scores. Note that it was *not* necessary to enter the F_{max} table with $df = n_{min} - 1 = 7$, since we are sure to accept at the smaller df value if we accept at the larger df value.

For the analysis of variance, the null hypothesis is $H_0: \mu_1 = \mu_2 = \mu_3 = \mu_4$. The computations leading to the sums of squares are:

Total sum of squares:

$$SS_T = \sum_{j=1}^{k} \sum_{i=1}^{n_j} X_{ij}^2 - \frac{X_{..}^2}{n_t} = 569 - \frac{123^2}{37} = 569 - 408.89 = 160.11$$

Treatment sum of squares:

$$SS_{treat} = \sum_{j=1}^{k} \frac{X_{.j}^2}{n_j} - \frac{X_{..}^2}{n_t} = \left(\frac{24^2}{9} + \frac{21^2}{10} + \frac{38^2}{8} + \frac{40^2}{10}\right) - \frac{123^2}{37}$$

$$= (64.0 + 44.1 + 180.5 + 160.0) - 408.89 = 448.60 - 408.89 = 39.71$$

Error sum of squares:

$$SS_{error} = SS_T - SS_{treat} = 160.11 - 39.71 = 120.40$$

In Table 6.6 these results are presented in an analysis-of-variance summary table.

TABLE 6.6   *Analysis-of-variance summary table*

Source	Degrees of freedom	Sum of squares	Mean square	F
Treatment	3	39.71	13.24	3.63
Error	33	120.40	3.65	
Total	36	160.11		

With 3 and 33 degrees of freedom, an F value of at least 2.90 is required for rejection of the null hypothesis with $\alpha = .05$. Hence, we reject the null hypothesis and conclude that significant differences exist among the sample arithmetic means. The interpretation of this result is that if the total population, of which our samples are representative, were run through one or another of the treatment conditions, we would find different arithmetic means for the subjects. However, this is a limited generalization of the results, since the original sample was not randomly selected but represented a volunteer group of students. Certainly no generalization of the freshman class as a whole is possible because it is very unlikely that volunteer subjects are truly representative of the total freshman class.

In order to broaden the interpretation of the experimental result, it would be necessary to replicate the experiment on other groups of subjects who represent a wider population (e.g., by including nonvolunteer subjects in a repetition of the experiment).

This experiment does illustrate the difference between *internal randomization* and *external randomization*. We have external randomization if the basic supply of experimental subjects is randomly selected from a wider population; we have internal randomization if a supply of subjects is randomly divided among treatment groups, whether or not the supply itself is randomly chosen. If an experiment has internal but not external randomization, we must, for all practical purposes, limit the generalization of results to the supply of subjects. (Although we might generalize to all groups of subjects "similar" to our supply, identifying such groups would be very difficult; for example, if subjects are volunteers, we do not know what characteristics are associated with volunteering for experiments). In general, our confidence in results from such experiments increases as additional replications (i.e., repetitions) of the experiment are carried out on new populations of subjects. The failure to plan replications of experiments is often advanced as an important criticism against educational research. The authors agree with this criticism and urge students to consider carefully the results when only internal randomization is used in experiments. The most convincing types of educational research studies are those in which replications on different populations are built into the experimental design.

Problems

6.1. Why is the arithmetic mean of an F distribution always greater than its median?

6.2. Using the table of percentiles of F distributions, find the relevant critical values for a test at the .01 level of significance when the hypothesis regards the equivalence of population variances and the sample sizes are:

  a. $n_1 = 5, n_2 = 10$
  b. $n_1 = 4, n_2 = 6$
  c. $n_1 = 9, n_2 = 9$

6.3. Using the table of percentiles of F distributions, find the relevant critical values for analysis of variance F ratios when the level of significance is .05 and the numbers of levels of treatment (k) and the samples sizes per treatment group (n) are:

  a. $k = 3, n = 10$
  b. $k = 5, n = 20$
  c. $k = 6, n = 2$

6.4. Two random samples of boys from two different physical education programs were tested on their ability to do pushups. The first sample contained 10 boys and the second sample, 16 boys.

Sample 1		Sample 2		
30	40	38	36	24
20	26	42	46	32
32	30	36	32	30
18	20	34	26	26
29	35	30	20	24
		28		

a. Compute the sample variances.

b. Test the data for homogeneity of variance using the .05 level of significance.

6.5.   Data are presented below for six random samples of subjects.   Using the F_{max} statistics, test for homogeneity of variance at the .05 level of significance. Do the unequal sample sizes lead to any ambiguity in the F_{max} test?

Sample	1	2	3	4	5	6
Size	7	8	4	10	15	8
Variance	14.21	17.50	5.25	8.32	6.67	11.12

6.6.   A driver-education teacher tested two random samples of boys on driving skills and obtained the results below.   Can the instructor conclude that the boys came from the same population with respect to their variability?   Use the .10 level of significance.

Sample 1		Sample 2	
5	7	14	12
15	9	10	16
5	5	12	10
8	11	13	13
12	7	11	12
10	11	8	16
4	6	7	13
7	6	15	10

6.7.   What is the difference between the F-ratio test used in the one-dimensional analysis of variance and the F-ratio test for homogeneity of variance?

6.8.   The following analysis-of-variance summary table was set up based on an experiment.

Source	Degrees of freedom	Sum of squares	Mean square
Treatment	2	97.42	48.71
Error	15	185.50	12.37
Total	17	282.92	

a. How many treatment levels were involved in the experiment?

b. Assuming equal group sizes, what was the number of subjects per group?

c. State the null hypothesis in symbolic form.

d. Would the null hypothesis be rejected at the .05 level of significance? At the .01 level of significance?

6.9. Randomly formed groups of potential school dropouts were randomly assigned to three counseling groups. After 6 months of individual counseling, they were given a personality test that yields scores on social adjustment. The data are presented below.

Group 1		Group 2		Group 3	
16	10	9	10	20	10
17	14	11	12	16	16
17	8	8	10	14	15
16	7	7	7	13	14
6	9	13	13	17	8

Complete an analysis of variance to decide if the treatments produced significantly different degrees of social adjustment. Use both the .05 and the .01 levels of significance.

6.10. The same professor taught four large sections of an elementary statistics course. At midterm, 10 students were randomly selected from each section and given the same examination. Based on the data presented below, do the sections differ significantly (at the .05 level) in achievement?

	Section		
1	2	3	4
38	38	45	30
33	31	33	38
39	29	23	34
31	33	41	38
22	37	35	26
40	16	33	36
37	20	24	26
32	33	15	23
29	20	46	37
26	20	41	21

Seven

methods for bivariate measured data

In Chap. 3, the concept of determining the degree of relationship between two variables for one group of subjects was explored for categorical data. When each of the two variables is measured on at least an equal-interval scale of measurement, there exists the greatest possibility for interpreting and utilizing in a predictive manner the information for the bivariate distribution of scores. In this chapter we begin by considering, in Sec. 7.1, descriptive techniques that can be applied to bivariate distributions. In Sec. 7.2, the notion of deriving equations for predicting values of one variable from values of another variable is developed. Section 7.3 presents an appropriate measure of correlation, as well as the way in which this measure of correlation is related to prediction. The final part of the chapter, Sec. 7.4, discusses testing hypotheses related to correlation measures.

7.1 Descriptive techniques

Bivariate interval or ratio data are generally displayed initially in a table of raw scores. Table 7.1 shows scores from an algebra pretest and a final statistics

TABLE 7.1   Distribution of pretest-algebra and final-statistics-examination scores for 28 beginning statistics students

Student number	Pretest score	Final score	Student number	Pretest score	Final score
1	42	89	15	40	83
2	29	73	16	39	81
3	36	87	17	31	81
4	24	68	18	26	62
5	32	80	19	33	78
6	35	84	20	27	74
7	46	91	21	38	80
8	38	90	22	40	74
9	22	70	23	34	69
10	27	64	24	35	78
11	34	86	25	31	71
12	20	63	26	37	66
13	38	84	27	28	80
14	41	87	28	29	72

examination for one group of beginning statistics students. This raw-score presentation does not reveal a great deal concerning how scores on these two examinations are related. However, if the various pairs of scores are studied, it can be seen that there is a tendency for high pretest scores to be associated with high final-examination scores and, similarly, for low scores to go together. In order to formalize this impression, the simple device of plotting each individual's pair of scores on ordinary rectangular, or Cartesian, coordinates is a very effective procedure. The horizontal axis is scaled in terms of the variable measured first in time, the pretest scores. On the vertical axis, the scale for final examination scores is shown. Figure 7.1 shows these coordinates for the

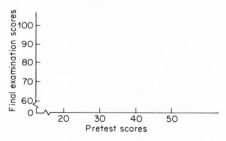

figure 7.1   Coordinate axes for a bivariate distribution of pretest and final-examination scores.

two tests. Note that the axes have been marked off to encompass only the values that occurred on each variable. That is, the horizontal axis was not started at 0 but at 20, since no score below 20 occurs in the pretest-score distribution. Similarly, the vertical axis was begun at 60 to include the smallest score, 62, in the final-examination-score distribution.

If the pairs of scores are used to graph a point for each subject, the resulting distribution of points is known as a *scatter diagram*. This name is appropriate because, with educational and psychological variables, one rarely finds that the points fall on any simple line or curve. Rather, the points are usually scattered about, although some trend may be more or less obvious in the distribution of points. The scatter diagram for the algebra- and statistics-test scores is shown in Fig. 7.2. This diagram confirms the previous observation concerning the relationship between the two types of scores; that is, high scores tend to go together, as do low scores.

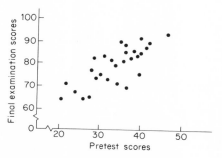

figure 7.2   *Scatter diagram for pretest and final-examination scores.*

The scatter diagram is the basic graphic procedure that will be applied to bivariate distributions. For this reason, a few comments concerning this procedure are especially relevant. (1) The horizontal axis is conventionally scaled in terms of either a variable measured first in time *or* a variable *from which* prediction is made if no clear-cut time sequence can be determined. This variable is known as the *predictor* variable and is symbolized as X. The vertical axis is scaled in terms of either a variable measured later in time *or* a variable it is desired to predict. Thus, this variable is known as the *predicted* variable and is symbolized as Y. An alternative terminology, not used in this text, refers to X as the *independent* variable and to Y as the *dependent* variable. (2) The units of the two variables need not be comparable, but they both must be on at least equal-interval scales of measurement. Thus X, for example, could be height measurements in inches for a group of students and Y could be their weights in pounds. It is perfectly meaningful to talk about relationships between different kinds of variables such as height and weight. (3) Once the axes are scaled, a point is graphed for each individual in the group. This point is located above the appropriate X value at a height corresponding to the Y value. Thus, in Figure 7.2, individual 1 is represented by a point located perpendicularly *above* 42 on the horizontal axis, since his pretest score was 42, and at a height corresponding to 89 on the vertical axis, since his final score was 89. Figure 7.3 illustrates, in detail, the graphing of this point. (4) Some-

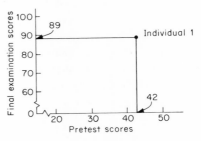

figure   7.3   *Graphing the point for*
student 1 from Table 7.1.

times two or more individuals may have identical scores on both variables. Since these persons are represented by exactly the same point on the scatter diagram, it is desirable to indicate the fact that the point stands for more than one individual. This can be done by the convention of placing a small numeral near the point on the scatter diagram; this numeral would be 2 if two individuals had identical scores; 3 if three individuals had identical scores; and so forth. (5) A scatter diagram should be carefully constructed by using a ruler or straightedge. A clear, plastic triangle is useful for graphing the points.

7.2   *Fitting prediction lines*

Discovery of functional relationships between two variables has both theoretical and practical origins. In a general sense, science has the goal of discovering and stating how one variable influences and is influenced by other variables. It is of interest from a scientific point of view, for example, to study the relationship between intelligence and such variables as school achievement, physical strength, parents' intelligence, and so forth in order to understand better the nature of intellect and to explain the role of intellect in a broader framework of human behavior. This interest in discovering functional relationships involving intelligence also has practical implications for such fields as education, politics, etc. From a practical viewpoint, prediction of one variable from some other variable has important applications in education. One common example involves predicting college success from high school grade-point average. Once the relationship between high school and college success has been determined, this information can be used selectively for future college applicants. Only these students with high school grade-point averages that are high enough to be predictive of more than minimal college success may be selected for admission.

In some situations, especially in the physical sciences, the relationship between variables can be discovered in a straightforward manner. If one variable, the X variable, is varied over a range of its values under standardized conditions, then the corresponding values of a second variable, the Y variable, can be observed and recorded. The relationship between X and Y, when observations are made under controlled, standardized conditions, can often be discovered by simply studying the pairs of X, Y scores. The simplest relationship between X and Y would occur when Y is a linear function of X. That is, if the observations were graphed on rectangular coordinates, the resulting graph would be a straight line. In general form, the mathematical equation for a straight line may be written as $Y = \alpha + \beta X$, where X and Y are values of the predictor and predicted variables, respectively, and α and β are numerical constants. Of course, X and Y are often related by some more complex function. If the relationship could be expressed by the formula $Y = \alpha + \beta X^2$, the resulting graph would be a parabola. Also, exponential and more complex relationships are not uncommon in the physical sciences.

Unfortunately, the behavioral scientist usually cannot control his observations or standardize the conditions of observation to the same degree that is possible in the physical sciences. For this reason, when he observes values for two variables, no clear-cut functional relationship may be readily apparent. Indeed, the kind of relationship illustrated for the test scores in Fig. 7.2 is typical of the data that arise in educational and psychological research. There is a suggestion in Fig. 7.2, of a functional relationship between algebra-pretest and statistics-final-examination scores, but the conditions of learning of the students and of testing are not controlled sufficiently to reveal an unambiguous relationship. However, it is often impossible, or at least impractical, in the behavioral sciences to exercise the type of control that would yield completely reliable observations.

Thus, the behavioral scientist is faced with imperfect data, but he has the same desire as the physical scientist to discover functional relationships among variables. In order to cope with this problem, certain reasonable assumptions are made concerning the type of relationships between the variables. In this textbook, we restrict ourselves to considering only the simplest case but the one most commonly utilized in educational research. The case is referred to as the *linear prediction model*. In effect, it is assumed that the two variables X and Y are related by a linear, or straight-line, equation *but* that various errors associated with the conditions of measurement, of the subjects, and so forth, also enter into the relationship. In mathematical form, we assume that Y is related to X by the equation $Y_i = \alpha + \beta X_i + \epsilon_i$. As before, X and Y are the predictor and predicted variables, respectively; also, α and β are numerical constants. The new term, ϵ_i, is an *error* term associated with the ith individual. Note that it is assumed that the error is additive; that is, ϵ_i is added to $\alpha + \beta X_i$ to obtain the Y value. Before explaining the linear prediction model in more detail, let us stress one important precaution. In practice, the model would be used only if one could reasonably assume that X and Y are, in fact, linearly related. If a scatter diagram is plotted and some nonlinear, say parabolic or exponential, trend is suggested by the data, the researcher would certainly not wish to attempt to force these data on the Procrustean bed of the linear prediction model. Instead, techniques appropriate for nonlinear relationships would be used.

Let us return to the linear prediction model. Without additional restriction, the model does *not* lead to useful prediction equations. In fact, if α and β are given any arbitrary values, the ϵ_i term can always be adjusted to yield perfect prediction of Y from X. For example, considering the data in Table 7.1, suppose that we arbitrarily let $\alpha = 10$ and $\beta = 1$. For individual 1, $X = 42$ and $Y = 89$. If $\epsilon_i = 37$, then $Y_1 = \alpha + \beta X_1 + \epsilon_1$, or $89 = 10 + 1(42) + 37$. Any other values could have been chosen for α and β and some appropriate value for ϵ_i would then give a perfect solution for the prediction equation. Obviously, some criterion must be established for the choice of α and β before the model is meaningful.

It is necessary at this point to digress slightly in order to consider the interpretation of the constants α and β in a linear equation. The general equation for a straight line is $Y = \alpha + \beta X$. α is numerically equal to the value of Y when $X = 0$. That is, α is referred to as the Y *intercept*. β is known as the *slope* of the line and is equal to the amount of change in Y for each 1-unit change in X. For example, if we consider the straight line

$$Y = 2 + 3X$$

this line crosses the Y axis at the point $Y = 2$; also, for each unit change in X, Y changes $+3$ units. Thus, for $X = 2$, $Y = 8$ and for $X = 3$, $Y = 11$. These interpretations of the constants α and β are shown graphically in Fig. 7.4.

Let us return now to the linear prediction model $Y_i = \alpha + \beta X_i + \epsilon_i$. The task remains of establishing some criterion for the selection of specific numerical values for α and β that will yield a unique and meaningful interpretation of the resulting prediction equation. Whereas other criteria have been proposed and used in certain applications, the most generally accepted criterion

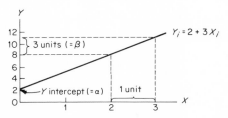

figure 7.4   *Graphical interpretation of α and β.*

can be stated as follows: The values of α and β are chosen in such a way that the average *squared* error is a minimum; that is,

$$\frac{\sum_{i=1}^{N} \epsilon_i^2}{N}$$

is as small as possible when we are fitting the prediction equation to a set of N points.   Since N is a constant, simply minimizing

$$\sum_{i=1}^{N} \epsilon_i^2$$

is sufficient and the condition is known as the *least-squares criterion*.

If we apply the least-squares criterion to the linear prediction model, we find a unique solution for α and β.   Only a brief outline of the solution will be presented here.   From the linear prediction model, $Y_i = \alpha + \beta X_i + \epsilon_i$; then, $\epsilon_i = Y_i - \alpha - \beta X_i$.   It can be shown that the least-squares criterion is met† if we choose an α value as $\mu_y - \beta\mu_x$ and a β value as

$$\frac{\sum_{i=1}^{N} X_i Y_i - N\mu_x\mu_y}{\sum_{i=1}^{N} X_i^2 - N\mu_x^2}$$

To distinguish these values from other α and β values, we shall use a "cap" on the constants and write

$$\hat{\alpha} = \mu_y - \hat{\beta}\mu_x \qquad \text{and} \qquad \hat{\beta} = \frac{\sum_{i=1}^{N} X_i Y_i - N\mu_x\mu_y}{\sum_{i=1}^{N} X_i^2 - N\mu_x^2}$$

† For the student who has been introduced to calculus, the procedure involves taking partial derivatives with respect to α and then with respect to β.   These are each set equal to 0 and solved simultaneously, yielding the solutions given above.   It is then easy to show that these are, indeed, minimum values.

Note that in order to compute α and β from an actual bivariate distribution of scores, it is necessary to find

$$\sum_{i=1}^{N} X_i$$

and, then, μ_x;

$$\sum_{i=1}^{N} Y_i$$

and, then, μ_y;

$$\sum_{i=1}^{N} X_i^2 \quad \text{and} \quad \sum_{i=1}^{N} X_i Y_i$$

The final term,

$$\sum_{i=1}^{N} X_i Y_i$$

is the only one not familiar to you from previous chapters. This term is simply the sum of the products of the X and Y scores per individual in the group.

If $\hat{\alpha}$ and $\hat{\beta}$ are used in the linear prediction model, the resulting average squared error,

$$\frac{\sum_{i=1}^{N} \epsilon_i^2}{N}$$

will be smaller than for any other choice of α and β. In this sense, using the least-squares values $\hat{\alpha}$ and $\hat{\beta}$ results in a prediction equation that guarantees the "best" prediction of Y from X.†

In actual prediction situations, once the values of $\hat{\alpha}$ and $\hat{\beta}$ have been computed from the data, a straight line can be constructed on the scatter diagram using these values. This straight line will be $\hat{Y}_i = \hat{\alpha} + \hat{\beta} X_i$. Note that $\hat{Y}_i$ is used in place of Y_i in this equation and that the term ϵ_i does not appear. $\hat{Y}_i$ is referred to as the *predicted Y value;* in general, $\hat{Y}_i$ and Y_i will have different values, unless Y_i happens to lie on the prediction line. These

† In this text, we always use Y to stand for the predicted variable. Thus, we always talk of predicting Y from X. *If* we reverse the roles of X and Y, the result for $\hat{\alpha}$ and $\hat{\beta}$ will, in general, be different. There is a different least-squares solution for predicting X from Y than for predicting Y from X. This distinction should be kept in mind so that the roles of the two variables are not confused.

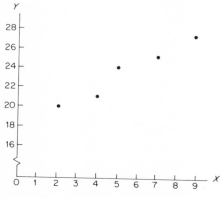

figure 7.5   Illustrative scatter diagram.

notions can be more clearly grasped by considering a contrived example with scores for just five individuals. Table 7.2 presents the data and the scatter diagram is shown in Fig. 7.5. There is an apparent linear trend in these data, although the points certainly do not fall exactly on a straight line. To solve for $\hat{\alpha}$ and $\hat{\beta}$ we need

$$\sum_{i=1}^{5} X_i = 27 \quad \text{and} \quad \mu_x = \frac{27}{5} = 5.4$$

$$\sum_{i=1}^{5} Y_i = 117 \quad \text{and} \quad \mu_y = 23.4$$

$$\sum_{i=1}^{5} X_i{}^2 = 5^2 + \cdots + 2^2 = 175 \quad \text{and}$$

$$\sum_{i=1}^{5} X_i Y_i = 5 \cdot 24 + 4 \cdot 21 + \cdots + 2 \cdot 20 = 662$$

TABLE 7.2   Illustrative bivariate distribution

Individual	X	Y
1	5	24
2	4	21
3	9	27
4	7	25
5	2	20

Solving for $\hat{\beta}$, we obtain

$$\hat{\beta} = \frac{662 - 5(5.4)(23.4)}{175 - 5(5.4)^2} = \frac{662 - 631.80}{175 - 145.80} = \frac{30.2}{29.2} = 1.03$$

Then, $\hat{\alpha} = 23.4 - (1.03)(5.4) = 23.4 - 5.56 = 17.84$.

The equation, $\hat{Y}_i = \hat{\alpha} + \hat{\beta}X_i$, or $\hat{Y}_i = 17.84 + 1.03X_i$, is shown superimposed on the scatter diagram in Fig. 7.6.

The prediction line always passes through the point of intersection of the arithmetic means (μ_x, μ_y).† This fact is useful in graphing the line. First locate a point (μ_x, μ_y) on the diagram. Then draw the line through this point intersecting the Y axis at the value $\hat{\alpha}$. If greater accuracy of graphing is desired, locate a third point near the high end of the X axis by solving the $\hat{Y}_i$ equation. By inspecting Fig. 7.6, it is apparent that the straight line does

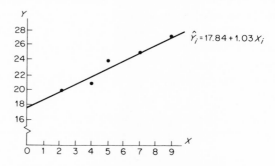

figure 7.6   *Scatter diagram and prediction line.*

not pass exactly through any of the points. The error term ϵ_i for a given individual is equivalent to the vertical distance from that individual's point on the scatter diagram to the prediction line. In Fig. 7.7, this is illustrated for individual 2, who had scores of $X_2 = 4$ and $Y_2 = 21$. When X is 4, the predicted Y value is $\hat{Y}_2 = 17.84 + 1.03(4) = 17.84 + 4.12 = 21.96$, and this point lies on the prediction line. The difference, $Y_2 - \hat{Y}_2$, is the error of prediction ϵ_2. In this case, $\epsilon_2 = -.96$, since $Y_2 = 21$ and $\hat{Y}_2 = 21.96$.

The value of the average squared error of prediction

$$\frac{\sum_{i=1}^{N} \epsilon_i^2}{N}$$

† This fact becomes apparent if we let $X_i = \mu_x$ in the prediction equation and then substitute $\mu_y - \beta\mu_x$ for α in the equation.

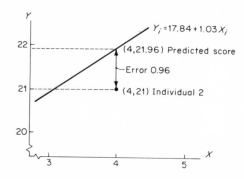

figure 7.7   Detail of prediction line for individual 2.

can be found for this example by computing ϵ_i for each individual and then squaring and summing these values. These steps are shown in Table 7.3. The sum of the squared errors

$$\sum_{i=1}^{5} \epsilon_i^2$$

is 1.96. The average squared error is, then, $1.96/5 = .39$.

TABLE 7.3   *Computation of the variance-of-estimate error*

Individual	X	Y	$\hat{Y}$	$(Y - \hat{Y}) = \epsilon$	$(Y - \hat{Y})^2 = \epsilon^2$
1	5	24	22.99	1.01	1.02
2	4	21	21.96	−.96	.92
3	9	27	27.11	−.11	.01
4	7	25	25.05	−.05	.00
5	2	20	19.90	.10	.01
					$1.96 = \sum_{i=1}^{5} \epsilon_i^2$

The average squared error

$$\frac{\sum_{i=1}^{N} \epsilon_i^2}{N}$$

is known, technically, as the *variance-of-estimate error* and is symbolized $\sigma_{y \cdot x}^2$. To explain this choice of name and symbol, let us consider the similarity between the variance-of-estimate error $\sigma_{y \cdot x}^2$ and the definition of the ordinary variance

$\sigma_y{}^2$ of a set of Y scores. Focusing only on the five Y scores in Table 7.3, we could compute their variance:

$$\sigma_y{}^2 = \frac{\sum_{i=1}^{N} (Y_i - \mu_y)^2}{N}$$

In the numerator, the arithmetic mean is subtracted from each score, and the difference is squared and then summed over all individuals in the group. On the other hand, the numerator of the variance-of-estimate error $\sigma_{y \cdot x}^2$ is found by subtracting the predicted Y score, $\hat{Y}_i$, from the corresponding Y score, squaring the difference, and summing over all individuals in the group. The only difference between the two sets of computations concerns the term subtracted from the Y scores. For the variance $\sigma_y{}^2$, the subtractive term is always $\mu_y{}^2$, for the variance-of-estimate error $\sigma_{y \cdot x}^2$, the subtractive term is not a constant but the predicted Y score, $\hat{Y}_i$. The $\hat{Y}_i$ score may be different for each different Y_i value.

The variance $\sigma_y{}^2$ is, in effect, a measure of variability around the arithmetic mean μ_y. By analogy, the variance-of-estimate error $\sigma_{y \cdot x}^2$ is a measure of variability around a prediction line. If $\sigma_{y \cdot x}^2$ is relatively large, the points in the scatter diagram lie relatively far from the prediction line; if $\sigma_{y \cdot x}^2$ is relatively small, the points lie near the prediction line. A further analogy is possible. The square root of a variance is a standard deviation; that is, $\sigma_y = \sqrt{\sigma_y{}^2}$. The square root of the variance-of-estimate error is known as the *standard error of estimate;* that is, $\sigma_{y \cdot x} = \sqrt{\sigma_{y \cdot x}^2}$. In Chap. 5 you learned that for normal distributions of scores, there are systematic relationships between proportions of scores above, below, or between various score points and the standard deviation of a set of scores. For example, in a normal distribution of scores, about 68 percent of the scores fall within an interval of one standard deviation on each side of the arithmetic mean. Similarly, in a bivariate distribution of scores, *if* the distribution of the error terms, ϵ_i, is normal, then about 68 percent of the scores will fall within an interval of one standard error of estimate on each side of the prediction line. Figure 7.8 illustrates this for a hypothetical scatter diagram. The solid line, labeled $\hat{Y}_i$, is the prediction line; the two dashed lines have been constructed at vertical distances of $\sigma_{y \cdot x}$ units on either side of the prediction line. Assuming that the distribution of the $Y_i - \hat{Y}_i = \epsilon_i$ values is normal, about 68 percent of the points lie between the two dashed lines. All other proportionate relationships associated with normal distributions would also be accurate for this example. Thus, about 5 percent of the points would lie more than 1.96 standard-error-of-estimate units from the prediction line, etc.

In the final paragraphs of this section, we apply the prediction theory just developed to the data presented previously in Table 7.1. Before doing so, how-

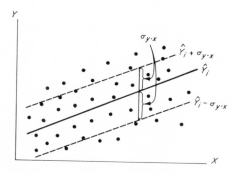

figure 7.8   *Scatter diagram with prediction line and standard-error-of-estimate lines.*

ever, we shall summarize the theory and procedures step by step.   (1) It is assumed that Y scores are linearly related to X scores but that there is error involved in the relationship.   Formally, the *linear prediction model* is written as $Y_i = \alpha + \beta X_i + \epsilon_i$.   (2) To obtain a unique and meaningful prediction line connecting X and Y, the *least-squares criterion* is applied.   Thus, the values $\hat{\alpha}$ and $\hat{\beta}$ are defined so that the average squared error

$$\frac{\sum_{i=1}^{N} \epsilon_i{}^2}{N}$$

is minimized (or, using different terminology, the value of the variance-of-estimate error $\sigma_{y.x}^2$ is minimized).   (3) The least-squares solutions for the prediction constants are

$$\hat{\alpha} = \mu_y - \hat{\beta}\mu_x \qquad \text{and} \qquad \hat{\beta} = \frac{\sum_{i=1}^{N} X_i Y_i - N\mu_x\mu_y}{\sum_{i=1}^{N} X_i{}^2 - N\mu_x{}^2}$$

The least-squares prediction equation is, then, $\hat{Y}_i = \hat{\alpha} + \hat{\beta}X_i$.   (4) For the ith X score, X_i, the corresponding actual Y score is Y_i.   The predicted Y score, from the prediction equation, is $\hat{Y}_i$.   The difference, $Y_i - \hat{Y}_i$, is the error ϵ_i.   The variability of the Y scores around the prediction line is expressed as the variance-of-estimate error and is the same as the average squared error; that is,

$$\sigma_{y.x}^2 = \frac{\sum_{i=1}^{N} (Y_i - \hat{Y}_i)^2}{N} = \frac{\sum_{i=1}^{N} \epsilon_i{}^2}{N}$$

(5) The square root of the variance-of-estimate error is known as the *standard error of estimate*. When the distribution of the ϵ_i values is normal, the proportionate relationships for normal distributions apply to the scatter of points around the prediction line.

We shall now apply these concepts to the algebra-pretest and statistics-final-examination data from Table 7.1. These data, in miniature, typify a common educational situation to which prediction equations have important application. Namely, it is common for one test, called a pretest, to be given to prospective students in order to predict their performance in specific courses of study. Thus, on entrance to college, a student may take an English comprehension aptitude test. Based on his score, his predicted success in beginning English composition is obtained. If the prediction is for poor or failing performance in the course, he may be placed in a remedial course prior to the composition course. One essential point should be noted. The prediction equation for such a selection purpose would be initially set up by administering the aptitude test to *all* entering students and then allowing *all* of them to complete the English composition course. Then, using (say) the final examination scores from the course as the predicted variable Y, and the aptitude scores as the predictor variable X, the least-squares prediction equation can be derived. Thereafter, this equation can be used for selection purposes with new students when only their aptitude scores are known. Of course, we must be sure that these new students are comparable to those on whom the prediction equation was initially developed. If changes in the entering student population occur, the prediction equation must be revised based on the new group.

The data in Table 7.1 can be used to build an equation for predicting success in a statistics course from scores on an algebra pretest. In real-life situation, this equation might subsequently be used to decide which students are "ready" for the statistics course and which need remedial instruction in mathematics. From Table 7.1, the necessary quantities can be found to compute the least-square prediction constants $\hat{\alpha}$ and $\hat{\beta}$. These terms are

$$\sum_{i=1}^{28} X_i = 932 \quad \text{and} \quad \mu_x = 33.29$$

$$\sum_{i=1}^{28} Y_i = 2,165 \quad \text{and} \quad \mu_y = 77.32$$

$$\sum_{i=1}^{28} X_i^2 = 32,136 \quad \text{and} \quad \sum_{i=1}^{28} X_i Y_i = 73,157$$

Then $\hat{\beta} = [73,157 - 28(33.29)(77.32)]/[32,136 - 28(33.29)^2] = .98$ and $\hat{\alpha} = 77.32 - .98(33.29) = 44.70$. The prediction equation is

$$\hat{Y}_i = 44.70 + .98X_i$$

This line is graphed on the scatter diagram (Fig. 7.9) by locating the point (μ_x,μ_y), or (33.29,77.32), and passing a line through this point and a second point that would also be on the line (for example, $X = 50$, $Y = 93.7$).

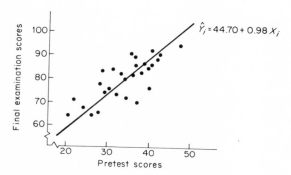

figure 7.9   *Prediction line for pretest and final-examination scores.*

The variance-of-estimate error and its square root, the standard error of estimate, can now be computed for this example by finding the error term, $Y_i - \hat{Y}_i$, for each of the 28 individuals in the distribution. For large sets of scores, however, this is a tedious procedure and shortcut techniques that avoid this tedium are available. The shortcut recommended at this point is based on the fact that the regression coefficient squared, $\hat{\beta}^2$, can also be expressed by the equation

$$\hat{\beta}^2 = \frac{\sigma_y{}^2 - \sigma_{y\cdot x}^2}{\sigma_x{}^2}$$

Solving for the variance-of-estimate error, we obtain $\sigma_{y\cdot x}^2 = \sigma_y{}^2 - \hat{\beta}^2\sigma_x{}^2$. Thus, the variance-of-estimate error can be found from the variances $\sigma_y{}^2$ and $\sigma_x{}^2$ and the slope $\hat{\beta}$. A second shortcut procedure for computing $\sigma_{y\cdot x}^2$ is presented in Sec. 7.3. Applying the formula to the present example requires finding $\sigma_y{}^2$, $\sigma_x{}^2$, and $\hat{\beta}^2$. Since $\hat{\beta} = .98$, $\hat{\beta}^2 = .98^2 = .96$. Using the raw scores in Table 8.1, we find that $\sigma_x{}^2 = 39.78$ and $\sigma_y{}^2 = 70.79$. Thus,

$$\sigma_{y\cdot x}^2 = 70.79 - .96(39.78) = 70.79 - 39.19 = 32.60$$

The standard error of estimate, the square root of $\sigma_{y\cdot x}^2$, is $\sqrt{32.60} = 5.71$.

All of the prediction concepts developed thus far have been presented in population notation. Fortunately, if only a random sample is available, very few changes in procedures are necessary. The sample arithmetic means are, of course, $\bar{X}$ and $\bar{Y}$. The sample variances are $S_x{}^2$ and $S_y{}^2$, with $n - 1$ in lieu of N, in the denominators of the formulas. For samples, we rewrite the prediction equation as $\hat{Y}_i = \hat{a} + \hat{b}X_i$, where $\hat{a}$ and $\hat{b}$ are estimates of $\hat{\alpha}$ and $\hat{\beta}$, respectively. Computationally, the sample estimates are found in exactly the same manner as the population parameter values. Thus, for a sample of n individuals,

$$\hat{b} = \frac{\sum\limits_{i=1}^{n} X_i Y_i - n\bar{X}\bar{Y}}{\sum\limits_{i=1}^{n} X_i^2 - n\bar{X}^2} \qquad \text{and} \qquad \hat{a} = \bar{Y} - \hat{b}\bar{X}$$

The major difference between the population and the sample computations occurs in the case of the variance-of-estimate error. To obtain an unbiased estimator of $\sigma_{y\cdot x}^2$, it is necessary to use $n - 2$ in the demoninator. Thus,

$$S_{y\cdot x}^2 = \frac{\sum\limits_{i=1}^{n} (Y_i - \hat{Y}_i)^2}{n - 2}$$

is the sample variance-of-estimate error. Similarly,

$$S_{y\cdot x} = \sqrt{S_{y\cdot x}^2} = \sqrt{\frac{\sum\limits_{i=1}^{n} (Y_i - \hat{Y}_i)^2}{n - 2}} = \sqrt{\left(\frac{n-1}{n-2}\right)(S_y^2 - \hat{b}^2 S_x^2)}$$

is the sample standard error of estimate. Since the sample computations are so similar to those previously illustrated for populations, no example will be considered.

7.3 Linear correlation

The variance-of-estimate error is a measure of the variability of the points in a scatter diagram about the least-squares prediction line. However, the variance-of-estimate error is in the units of the Y variable. Thus, if Y were weight measured in, say, pounds, the variance-of-estimate error $\sigma_{y\cdot x}^2$ would be in pound squared units. As an example, if $\sigma_{y\cdot x}^2 = 50$, this indicates 50 pounds squared. If the measurement had been in ounces instead of pounds, the variance-of-estimate error would have been 12,800 ounces squared rather than 50 pounds squared. Also, the standard error of estimate would be $\sqrt{50} = 7.07$ pounds, or $\sqrt{12,800} = 113.14$ ounces. The magnitude of either the variance-of-estimate error or its square root, the standard error of estimate, is dependent upon the choice of unit for Y variable. In the example, when the unit is ounces, the standard error of estimate is 16 times as large as when the unit is pounds. This dependence upon the choice of measurement unit makes both the variance-of-estimate error and the standard error of estimate difficult to interpret directly. If a prediction line is fitted by the least-squares method, we are generally interested in how well the prediction line fits the points in the scatter diagram. That is, we want to know if the variance-of-estimate error is *relatively* large or *relatively* small. Just examining the magnitude of the variance-of-estimate error will not tell us how well the prediction line fits, since the choice of unit for the Y variable alone can make $\sigma_{y \cdot x}^2$ large or small in magnitude. What is needed is some basis of comparison for $\sigma_{y \cdot x}^2$ so that its relative magnitude may be assessed. It turns out that an appropriate comparison can be made between $\sigma_{y \cdot x}^2$ and σ_y^2. This is true because the *largest* value that $\sigma_{y \cdot x}^2$, in theory, can assume is σ_y^2 for any bivariate distribution. An intuitive understanding of this relationship is possible from studying the three scatter diagrams in Fig. 7.10. In each

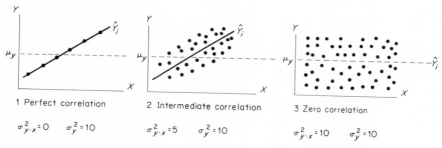

1 Perfect correlation $\qquad$ 2 Intermediate correlation $\qquad$ 3 Zero correlation

$\sigma_{y \cdot x}^2 = 0 \qquad \sigma_y^2 = 10 \qquad\qquad \sigma_{y \cdot x}^2 = 5 \qquad \sigma_y^2 = 10 \qquad\qquad \sigma_{y \cdot x}^2 = 10 \qquad \sigma_y^2 = 10$

figure 7.10   Scatter diagrams illustrating three degrees of relationship.

example, the variance of the Y scores is assumed to be 10. The variance is, of course, a measure of variability about the μ_y value. In the first example, the points fall perfectly on the prediction line and the variables X and Y are perfectly correlated. Since there is no scatter of points about the prediction line, the variance-of-estimate error is 0. In this example, however, the variance about μ_y is still 10. The prediction line in the second example is not a perfect fit to the points, although there is a definite linear relationship between X and Y. If $\sigma_{y \cdot x}^2$ is computed for this example, it is, say, 5. If you study example 2, it is evident that $\sigma_{y \cdot x}^2$ will be less than σ_y^2, since the points are less disperse about the prediction line than they are about the mean line μ_y. In the third example, X and Y are completely uncorrelated; in this situation, if $\hat{\beta}$ is computed it will be found to equal 0. Thus, the prediction line is $\hat{Y}_i = \hat{\alpha} + \hat{\beta} X_i = \hat{\alpha} + 0 X_i$, or $\hat{Y}_i = \hat{\alpha}$. But, $\hat{\alpha} = \mu_y - \hat{\beta} \mu_x = \mu_y - 0 \mu_y$, or $\hat{\alpha} = \mu_y$. Thus, $\hat{Y}_i = \mu_y$ is the prediction line when X and Y are unrelated. Interpretatively, this means that when X and Y are uncorrelated, knowing X does not help you predict Y at all. In this case, the best prediction for Y is always its arithmetic mean μ_y. Using $\hat{Y}_i = \mu_y$ as the prediction line results in $\sigma_{y \cdot x}^2 = \sigma_y^2$. The student can verify this easily by writing the formula for $\sigma_{y \cdot x}^2$ and then substituting μ_y for $\hat{Y}_i$ in the formula. The resulting formula is identical to that for the variance σ_y^2. In summary, the variance-of-estimate error is 0 when perfect prediction exists; it is the same as σ_y^2 when Y cannot be predicted from X; and it takes on values between 0 and σ_y^2 for intermediate degrees of relationship between X and Y.

The relationship between $\sigma_{y \cdot x}^2$ and σ_y^2 can be used to define a very important measure of the degree of correlation between X and Y. This measure is also related to how well the prediction line fits the points in the scatter diagram. Let us pursue the chain of reasoning that results in the definition of this important derived measure.

If the ratio of the variance-of-estimate error to the variance of Y scores is formed, we obtain $\sigma_{y \cdot x}^2/\sigma_y^2$. The range of this ratio is from 0 to 1 because (1) when $\sigma_{y \cdot x}^2$ is at its minimum, which is 0, the ratio is $0/\sigma_y^2$ or 0; and (2) when $\sigma_{y \cdot x}^2$ is at its maximum, which is σ_y^2, the ratio is σ_y^2/σ_y^2, or 1. Thus, $\sigma_{y \cdot x}^2/\sigma_y^2$ is 0 for perfect correlation between X and Y and is 1 for no correlation between X and Y. Note that in Fig. 7.10, example 1, $\sigma_{y \cdot x}^2/\sigma_y^2$ is 0/10 and that X and Y are perfectly related in this example; also, in example 3, X and Y are completely unrelated and $\sigma_{y \cdot x}^2/\sigma_y^2$ is 10/10 or 1. However, in previous discussions of correlational measures, it was true that 0 meant no correlation and 1 meant perfect correlation. To make the present measure conform to previous usage, a simple trick is utilized. Instead of the ratio $\sigma_{y \cdot x}^2/\sigma_y^2$, we define our correlation measure as $1 - (\sigma_{y \cdot x}^2/\sigma_y^2)$. Now, when $\sigma_{y \cdot x}^2/\sigma_y^2 = 0$, $1 - 0 = 1$; and when $\sigma_{y \cdot x}^2/\sigma_y^2 = 1$, $1 - 1 = 0$. Thus, the scale has been reversed.

The quantity $1 - (\sigma_{y \cdot x}^2/\sigma_y^2)$ is known as the *coefficient of determination* and is symbolized as ρ^2. The coefficient of determination can be interpreted as either a measure of the predictive worth of X as a predictor of Y or as a correlational measure. These two interpretations and their basic equivalence are discussed below in detail.

For interpretive purposes, the formula for the coefficient of determination can be rewritten as $\rho^2 = (\sigma_y^2 - \sigma_{y \cdot x}^2)/\sigma_y^2$. The numerator is the difference between σ_y^2 and $\sigma_{y \cdot x}^2$. It is, therefore, the amount by which $\sigma_{y \cdot x}^2$ is smaller than σ_y^2. In effect, $\sigma_y^2 - \sigma_{y \cdot x}^2$ is the reduction in variability in Y scores as a result of the relationship between X and Y. This reduction in variability occurs because, in general, the scores deviate less from the prediction line than they do from the arithmetic mean of the Y scores. The difference, $\sigma_y^2 - \sigma_{y \cdot x}^2$, is often referred to as the "explained" variability in Y, since it is the variability of the Y scores predictable from the X scores. Using this terminology, we say that the ratio $(\sigma_y{}^2 - \sigma_{y \cdot x}^2)/\sigma_y{}^2$ is the proportion of "explained" variability as compared to total variability $\sigma_y{}^2$. Since this ratio is equivalent to the coefficient of determination, ρ^2 represents the proportion of variability in Y scores predictable from, or explained by, X scores. By the same token, ρ^2 is a measure of the relationship, or correlation, between X and Y. The greater the degree to which the variability in Y scores can be explained by difference in X scores, the more highly correlated are X and Y. The extreme case occurs when all the Y variability can be explained by X scores alone, and we say that X and Y are perfectly correlated, or, alternately, that Y is completely predictable from X.

To illustrate these ideas, let us suppose that a college aptitude test has a coefficient of determination of .40 when the predicted variable Y is freshmen grade-point averages. This means that the use of the aptitude test as a predictor results in a variance-of-estimate error that is 40 percent less than the variance in grade-point average. On the other hand, the predicted Y scores still show 60 percent as much variability about the prediction line as the Y scores show variability about the arithmetic mean of the grade-point-average distribution. These notions are extremely important when assessing the practicability of utilizing specific predictors. From the point of view of cost, time, convenience, etc., a decision to use a predictor may be based on its ability to "explain" the variability in Y scores. A small gain, such as 5 or 10 percent in explained variability, may not be worthwhile in a practical situation. That is, if ρ^2 were only .05 or .10, it might not be efficient to use the predictor at all. Of course, such a decision depends upon the details of the prediction situation.

As a measure of the correlation between two variables, X and Y, the coefficient of determination ρ^2 has the disadvantage of not indicating the direction of correlation between X and Y. For example, the two scatter diagrams in Fig. 7.11 both show the same degree of relationship between the variables; but in the first case, the relationship is a direct, positive association, whereas in the second case, it is an inverse, negative relationship. The essential difference between the two diagrams is that the slope of the prediction line in the first case is positive, whereas it is negative in the second case. The coefficient

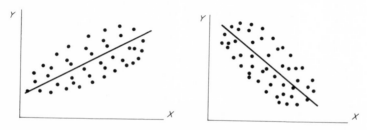

figure 7.11   Positive and negative correlations.

of determination ρ^2 is always positive. Since, from a practical point of view, it is convenient to have the sign on the correlational measure, the conventional measure of correlation for interval and ratio data is the square root of the coefficient of determination, with the sign taken from the slope $\hat{\beta}$ of the prediction line. Thus, if β is positive, we use $\sqrt{\rho^2} = \rho$; and if β is negative, we use $-\sqrt{\rho^2} = \rho$. For Fig. 7.11, we assume that $\rho^2 = .64$ in each case. Then, for the first diagram, the measure is $\rho = \sqrt{.64} = .80$, but for the second diagram, it is $\rho = -\sqrt{.64} = -.80$. Later, when we consider a computational formula for ρ, the formula will supply the proper sign automatically.

The correlational measure ρ is known as the *Pearson product-moment correlation coefficient*, although this name is frequently shortened to product-moment correlation, or Pearsonian correlation coefficient. In reports of educational research, such as those found in educational research journals, the correlation coefficient typically used is the Pearson product-moment correlation coefficient. When such coefficients are interpreted, it is important to remember that ρ^2 (not ρ) is directly related to the predictive worth of a variable. Thus, an aptitude test may report a validity coefficient of .50 for performance in a particular kind of course. The explained variability in achievement is, however, $.50^2$, or 25 percent, of the total variability, *not* 50 percent.

Let us turn now to a consideration of the mechanics of computing a Pearson product-moment correlation coefficient. To proceed from the definitional formula $\rho = \sqrt{1 - (\sigma_{y \cdot x}^2 / \sigma_y^2)}$ would be tedious in practice, since the computation of $\sigma_{y \cdot x}^2$ is lengthy. A shortcut procedure should always be used in practice. A formula, algebraically equivalent to the above, is

$$\rho = \frac{\sum_{i=1}^{N} X_i Y_i - N \mu_x \mu_y}{\sqrt{\left(\sum_{i=1}^{N} X_i^2 - N \mu_x^2\right)\left(\sum_{i=1}^{N} Y_i^2 - N \mu_y^2\right)}}$$

If the standard deviations σ_x and σ_y have already been computed, a more expedient formula is

$$\rho = \frac{\sum_{i=1}^{N} X_i Y_i - N \mu_x \mu_y}{N \sigma_x \sigma_y}$$

Since one generally wishes to compute the variances and standard deviations for both variables in a bivariate distribution of scores, the second formula will be used in this textbook. A variety of other computational formulas is available, all of them algebraically equivalent to those presented here. To determine ρ using our computational approach, first find the arithmetic mean and the standard deviation for both the X and Y scores. The only remaining term required in the formula is

$$\sum_{i=1}^{N} X_i Y_i$$

This is the same term that appears in the formula for the regression coefficient $\hat{\beta}$ (note that the numerator of the formula for $\hat{\beta}$ is identical to that of the formula for ρ). Once these terms are found, computation of ρ involves only substitution in the formula. The sign of ρ will be the same as the sign of $\hat{\beta}$ and, hence, the same as the slope of the prediction line. If the coefficient of determination ρ^2 is desired rather than ρ, proceed as above to find ρ, then square the result to obtain ρ^2.

We referred earlier to a second computational approach to the variance-of-estimate error $\sigma_{y \cdot x}^2$. The rationale for this approach is now clear. Since $\rho^2 = 1 - (\sigma_{y \cdot x}^2 / \sigma_y^2)$, it is easy to show algebraically that $\sigma_{y \cdot x}^2 = \sigma_y^2 (1 - \rho^2)$. Also, the standard error of estimate is $\sigma_{y \cdot x} = \sigma_y \sqrt{1 - \rho^2}$. Thus, the computational routine is to first compute ρ by the computational formula given above, square the result to obtain ρ^2, and then substitute to find $\sigma_{y \cdot x}^2$ or $\sigma_{y \cdot x}$.

A useful relationship between ρ and $\hat{\beta}$ is that $\rho = \hat{\beta}(\sigma_x / \sigma_y)$. This relationship is derived most easily by noting the similarities between the computational formulas for ρ and $\hat{\beta}$. The interested student can carry out this derivation with a little algebra if he rewrites the formula for $\hat{\beta}$ as

$$\frac{\sum_{i=1}^{N} X_i Y_i - N\mu_x \mu_y}{N\sigma_x^2}$$

The formula $\rho = \hat{\beta}(\sigma_x / \sigma_y)$ may be used computationally to find ρ if one has already computed $\hat{\beta}$ from the data.

If random samples rather than entire populations are involved in a research study, the formulas are changed only slightly. For inferential purposes, the computational formula for the sample Pearson product-moment correlation coefficient is

$$r = \frac{\sum_{i=1}^{n} X_i Y_i - n\bar{X}\bar{Y}}{(n-1)S_x S_y}$$

Note that sample estimates are used throughout, that N is replaced by n as the upper limit of the summation operator, that $n - 1$ replaces N in the denominator, and that the symbol r is used in lieu of ρ. Also, r^2 is the sample estimate of the coefficient of determination and may be used interpretatively in the same manner as ρ^2.

To compute the sample variance-of-estimate error, the formula is

$$S^2_{y \cdot x} = S_y{}^2 \frac{(n-1)}{(n-2)} (1 - r^2)$$

and the sample standard error of estimate is

$$S_{y \cdot x} = S_y \sqrt{\frac{n-1}{n-2} (1 - r^2)}$$

The factor $(n-1)/(n-2)$ in these formulas is necessary, since the denominator of $S_{y \cdot x}$ from the definitional formula is $n-2$, not $n-1$.

We previously derived the least-squares prediction line $\hat{Y}_i = 44.70 + .98X_i$ for the bivariate distribution of test scores from Table 7.1. Now we shall solve for the correlation between these scores and also determine the value of the variance-of-estimate error. Since these scores are assumed to represent an entire population, the parametric notation and formulas will be used. To compute ρ, we need the following terms (computed from the scores in Table 7.1):

$$\sum_{i=1}^{28} X_i = 932 \qquad \sum_{i=1}^{28} Y_i = 2{,}165$$

$$\sum_{i=1}^{28} X_i{}^2 = 32{,}136 \qquad \sum_{i=1}^{28} Y_i{}^2 = 169{,}383$$

$$\mu_x = \frac{932}{28} = 33.29 \qquad \mu_y = \frac{2{,}165}{28} = 77.32$$

$$\sigma_x = \sqrt{\frac{1{,}105.73}{28}} = \sqrt{39.49} = 6.28 \qquad \sigma_y = \sqrt{\frac{1{,}982.11}{28}} = \sqrt{70.79} = 8.41$$

$$\sum_{i=1}^{28} X_i Y_i = 73{,}157$$

Then

$$\rho = \frac{\sum_{i=1}^{N} X_i Y_i - N\mu_x\mu_y}{N\sigma_x\sigma_y} = \frac{73{,}157 - 28(33.29)(77.32)}{28(6.28)(8.41)} = \frac{1{,}085.48}{1{,}478.81} = .73$$

Also, the coefficient of determination is $\rho^2 = .73^2 = .53$. Thus, the variability around the prediction line is 53 percent *less* than the total variability $\sigma_y{}^2$ of the Y scores. Since $\hat{\beta}$ was already known to be .98, a somewhat simpler computational approach would have been to utilize the relationship $\rho = \beta(\sigma_x/\sigma_y)$. Thus, $\rho = .98(6.28/8.41) = .98(.75) = .74$. This value, within a small rounding error, agrees with our previous result.

Turning now to the variance-of-estimate error, we obtain

$$\sigma_{y.x}^2 = \sigma_y^2(1 - \rho^2) = 70.79(1 - .73^2) = 70.79(.47) = 33.27$$

Thus, the average squared error,

$$\frac{\sum_{i=1}^{N} \epsilon_i^2}{N}$$

about the prediction line is 33.27. The square root of 33.27, or 5.77, is the standard error of estimate $\sigma_{y.x}$. This result agrees, within an error due to rounding, with the result previously found.

7.4   *Testing hypotheses concerning correlations*

If a sample of n individuals is randomly selected and bivariate interval or ratio measurements are obtained, the procedures of the previous sections can be used to *estimate* the population-prediction equation, variance-of-estimate error, and correlation coefficient for the two variables. Since these estimated values are affected by chance factors related to sampling variability, we are aware, of course, that the sample estimates may deviate more or less from the actual-population values. In this section, we consider testing hypotheses in two situations. In the first case, the hypothesis of interest concerns the existence or nonexistence of an association between the two variables in the population. That is, we hypothesize that the population correlation between the two variables is 0 and then use the sample correlation coefficient to test the reasonableness of this assumption. In the second situation, we have two different random samples and wish to decide whether or not the populations yielding these samples have equal correlations between the two variables of interest. We begin by considering the first situation in detail.

If a random sample is measured along two variables and the correlation between these variables is computed and found to have the value, for example, $r = .30$ (where r is the sample estimate of ρ), it is possible that the population from which the sample was drawn shows no relationship between the two variables (that is, $\rho = 0$), but that $r = .30$ is due to sampling variability. Thus, if many additional samples were drawn from this population and r computed for each, we would find, in the long run, that r averaged out to 0. On the other hand, the value of $r = .30$ may be indicative of some real positive correlation between the two variables, and if this is true, .30 is the best available estimate of the value of ρ. When bivariate data are being analyzed, the researcher does not wish to draw any implications of practical significance until the question of the existence or nonexistence of a relationship has been settled. Hence, a relevant null hypothesis is H_0: $\rho = 0$. *If* the null hypothesis is true, then the sampling distribution of r will have its arithmetic mean at 0. This is equivalent to saying that in the long run, r would average out to 0 if repeated random sampling were conducted. When the null hypothesis is true, both positive and negative values of r will arise in different samples and the sampling distribution of r will be symmetric about the 0 value. In fact, the null hypothesis H_0: $\rho = 0$ can be tested by use of the already familiar t distributions. Since the arithmetic mean of the sampling distribution is assumed to be 0, t can be computed if some estimate of the standard deviation of this distribution is available. It turns out that when $\rho = 0$, the standard deviation of the sampling distribution of r is estimated by $\sqrt{(1 - r^2)/(n - 2)}$. Thus, the appropriate test statistic is

$$t = \frac{r - \rho}{\sqrt{(1 - r^2)/(n - 2)}}$$

Since the hypothesized value of ρ is 0, this is more conveniently written as $t = (r \sqrt{n - 2})/(\sqrt{1 - r^2})$. The degrees of freedom for this t test are $n - 2$. Note that the formula utilizes only r and n for the computation. If, for example, $r = .30$ were found between two variables measured on a sample of 25 individuals, the value of t would be

$$t = \frac{(.30) \sqrt{25 - 2}}{\sqrt{1 - .30^2}} = \frac{.30(4.80)}{\sqrt{.91}} = \frac{1.44}{.95} = 1.52$$

With $25 - 2 = 23$ degrees of freedom, and a nondirectional alternative (that is, $\rho \neq 0$), this value fails to reach significance. Hence, we conclude that when $n = 25$, $r = .30$ is not an unusual sample value for r, when, in fact, $\rho = 0$.

This test is very easy to use once r has been computed from a sample. The procedure can be simplified even further, however. Since the formula for t depends only upon r and n, it is possible to compute, for different degrees of freedom, the smallest values of r that would result in rejection of the null hypothesis at a given level of significance. To illustrate this approach, consider the general case for 23 degrees of freedom (that is, when $n = 25$). Going to the t table with 23 degrees of freedom, we find that at $\alpha = .05$, nondirectional test, the critical value of t is 2.07. Thus, to reject the null hypothesis H_0: $\rho = 0$, r must be large enough so that the resulting t is at least 2.07. If this is done, the value of r is .40. If the value of r from a sample of 25 cases is at least .40, then the null hypothesis can be rejected; otherwise it is accepted. Table F.7 presents a large number of "smallest significant r values" such as the one just computed. This table was built by systematically solving for r for various combinations of degrees of freedom and levels of significance. Both directional and nondirectional critical values are presented. To use Table F.7 in a practical situation, it is necessary only to compute r; then look in the appropriate degrees of freedom row $(df = n - 2)$ to see if the computed value of r exceeds the critical value. If it does, reject the null hypothesis; otherwise accept the null hypothesis.

The t test as presented above, or Table F.7, is only appropriate when the null hypothesis is $H_0: \rho = 0$. If one wishes to test against some other population value (say, $\rho = .50$), a problem arises. Under the assumption that $\rho = .50$, for example, the sampling distribution of r is no longer a symmetric distribution. This occurs because r cannot be larger than 1; hence, there is a kind of truncation of the sampling distribution. The arithmetic mean of the sampling distribution is still ρ, but the symmetry is destroyed and the t test is not an accurate or correct procedure. The explanation for the skewness of the sampling distribution of r when ρ is not 0 can be observed from the diagrams in Fig. 7.12. The

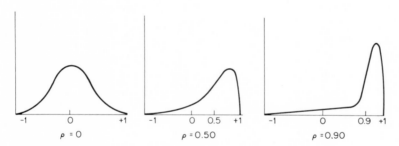

figure 7.12   Skewness of the sampling distribution of r for various values of ρ.

first diagram shows the symmetric case; that is, $\rho = 0$. The second diagram shows the negative skewness when $\rho = .50$. The third diagram displays the extreme skewness that occurs for large positive (or negative) values of ρ (for example $\rho = .90$). Since r cannot be larger than 1 (or smaller than -1), the sampling distribution "piles up" near 1 for cases when ρ is near 1 (and near -1 when ρ is near -1).

In order to cope with the problem arising from the skewness of the sampling distribution of r when ρ is not 0, the British statistician R. A. Fisher suggested the use of a transformation of r that results in normal sampling distributions. The transformation is $Z' = \frac{1}{2} \ln \left[(1 + r)/(1 - r) \right]$ where ln refers to a natural (Naperian) logarithm (i.e., the logarithm taken to base $e = 2.71828 \ldots$). The sampling distribution of Z' tends to be a normal distribution with arithmetic mean equal to ρ and standard deviation of $\sqrt{1/(n - 3)}$ for any value of ρ. Since working with logarithms may be unfamiliar to some readers of this textbook, Table F.8 provides a direct transformation from r to Z'. It is merely necessary to find r in the appropriate row and column of Table F.8 and then read off the corresponding value of Z'. For the null hypothesis H_0: $\rho = R$, where R is some numerical value assumed for ρ, we can use a z test:

$$z = \frac{Z' - Z_0'}{\sqrt{1/(n - 3)}}$$

and refer to the unit normal curve tables in order to decide on acceptance or rejection of the null hypothesis. The term Z_0' in the formula is the Z' value corresponding to R, the hypothesized value of ρ.

To illustrate, let us suppose that the null hypothesis H_0: $\rho = .50$ is relevant in a particular research situation. A sample of size 25 is drawn and the sample Pearson product-moment correlation coefficient is found to be $r = .38$. Thus, we wish to determine if $r = .38$ is an unusual sample value for r when $\rho = .50$ is the population value. Since the hypothesized value of ρ is not 0, we know that the sampling distribution of r is skewed. Hence, the use of the Fisher Z' transformation is indicated. From Table F.8, the value $Z' = .40$ is found for $r = .38$ and $Z_0' = .55$ is found for .50. Since $n = 25$, the test statistic is $z = (.40 - .55)/\sqrt{1/(25 - 3)} = -.15/.21 = -.71$. For a nondirectional test at $\alpha = .05$, the tabular z value is -1.96. Hence, the null hypothesis must be accepted, and we decide that based on a sample of 25 cases, $r = .38$ is not an unusual value when the population correlation is .50.

As mentioned at the beginning of this section, the second situation for which hypotheses may be tested concerns the difference of the correlations between the same two variables in two different populations. Thus, random samples are available from two populations, and it is of interest to decide whether or not these two populations have equal correlations between a specific pair of variables. For example, if scores from an aptitude test and measures of course achievement are correlated for a sample of males and also for a sample of females, it would be of interest to decide whether or not these two variables are equally correlated for males and females. If we let ρ_1 be the correlation in population 1 and ρ_2 the correlation in population 2, then the null hypothesis may be written as H_0: $\rho_1 = \rho_2$. The corresponding samples are of size n_1 and n_2, respectively. Also, the Pearsonian correlation coefficient can be computed from each sample. We call these r_1 and r_2, respectively. In order to test the null hypothesis, we use the sample information. To devise a test, we consider the sampling distribution of differences, $r_1 - r_2$, assuming that $\rho_1 = \rho_2$. The arithmetic mean of the sampling distribution is, of course, 0, since $\rho_1 - \rho_2 = 0$. However, unless ρ_1 and ρ_2 are each 0, the sampling distribution is skewed for the reasons just elaborated when discussing the one-sample test. Once again, fortunately, the use of the Fisher Z' transformation will result in a normal sampling distribution, and we may base the test of the null hypothesis on these Z' values rather than on the correlations themselves. Using Table F.8, we convert r_1 to Z'_1 and r_2 to Z'_2. The sampling distribution of the difference, $Z'_1 - Z'_2$, has a standard deviation of $\sqrt{1/(n_1 - 3) + 1/(n_2 - 3)}$. Since the sampling distribution is normal, an appropriate test statistic is

$$z = \frac{Z'_1 - Z'_2}{\sqrt{\dfrac{1}{n_1 - 3} + \dfrac{1}{n_2 - 3}}}$$

The computed z value can then be evaluated according to the unit normal curve tables. In general, a nondirectional test is relevant.

As an example of the two-sample z test for correlation coefficients, let us pursue the case of correlating aptitude and course achievement for a sample of males and a sample of females. Assume that $n_1 = 30$ males are randomly sampled and the Pearson product-moment correlation coefficient is found to be $r_1 = .47$. Similarly, for $n_2 = 50$ females, $r_2 = .53$. The research question of interest concerns the comparison of degree of correlation in the population of males and the population of females. Thus, the null hypothesis is $H_0: \rho_1 = \rho_2$ and a nondirectional test is appropriate. Using Table F.8, we find that for $r_1 = .47$, the corresponding Z' value is $Z'_1 = .51$ and for r_2, $Z'_2 = .59$. Substituting in the formula for z, we obtain

$$z = \frac{.51 - .59}{\sqrt{1/(30 - 3) + 1/(50 - 3)}} = -\frac{.08}{.24} = -.33$$

This computed value does not exceed the .05-level-of-significance critical value -1.96 for z; hence, the null hypothesis is accepted. We decide that the two populations are homogeneous with respect to their correlations.

Problems

7.1. Construct a scatter diagram for the following scores.

Student number	Pretest score	Final score
1	10	15
2	20	25
3	30	35
4	40	45
5	50	55

7.2.   An academic aptitude test was administered to a group of 10 children; their chronological ages and test scores were as follows.

Student number	Chronological age, months (X)	Aptitude test score (Y)
1	109	62
2	112	65
3	123	79
4	126	78
5	128	88
6	132	92
7	134	81
8	135	88
9	139	97
10	162	110

    a. From the scatter diagram for X and Y, does there appear to be a linear relationship?

    b. Determine the prediction equation.

    c. If $X = 142$, find the predicted value of Y from the prediction equation.

    d. Graph the prediction equation on the scatter diagram.

    e. Calculate r and test r for significance using $\alpha = .05$

    f. Calculate the sample standard error of estimate.

7.3.   The correlation between two variables is .87 and $S_X = 3.7$, $S_Y = 4.1$, $\bar{X} = 15.0$, $\bar{Y} = 32.0$.

    a. Determine the prediction equation for predicting Y from X.

    b. Find the predicted Y score for an X score of 27.

7.4.   For a correlation coefficient of .40, what proportion of the variance in Y is accounted for by X?   What proportion remains unaccounted for?   How much less than the standard deviation of Y is the standard error of estimate?

7.5.   Below are data for two variables.

Student	Variable 1	Variable 2
1	1	3
2	4	7
3	6	6
4	7	9
5	8	9
6	10	14

    a. Let variable 2 be the criterion variable and variable 1 be the predictor variable; find the prediction equation relating the two variables.

b. Let variable 1 be the criterion variable and variable 2 be the predictor variable; find the prediction equation relating the two variables.

c. Plot the scatter diagram and construct on it the two prediction equations from (a) and (b) above.

7.6. The correlation coefficient for scores on two standardized achievement tests, test X and test Y, is reported to be .427. Find the prediction equation if $S_X{}^2 = 239.9$, $S_Y{}^2 = 111.4$, $\bar{X} = 59.7$, and $\bar{Y} = 96.7$

7.7. A random sample of 60 students from elementary statistics classes at a large university revealed a correlation coefficient of .60 between overall grade-point average and final-examination scores in the statistics course. Decide whether or not it is reasonable to conclude that the population from which the sample was drawn shows no correlation between these two variables.

7.8. Using the data from Prob. 7.7., test the hypothesis that the population correlation between overall grade-point average and final-examination scores in statistics is .80; use the .01 level of significance.

7.9. Aptitude-test scores and scores from an achievement test were correlated for a sample of boys and for a sample of girls. The sample of boys contained 39 subjects and the correlation coefficient was .500; the sample of girls contained 67 subjects and the correlation coefficient was .753. Test the significance of the difference between these coefficients at the .05 level.

Eight

descriptive and inferential methods for rank-order data

Rank-order data based on ordinal scales of measurement generally arise in one of three ways. (1) Ranks are obtained directly for the members of a group (e.g., a teacher ranks her students on the degree of creativity that she believes each possesses; thus, she gives rank 1 to the student she believes is most creative, rank 2 to the student she believes is second most creative, and so forth). (2) Members of a group are classified into two or more categories that represent a rank order (e.g., school children are classified as under achievers, average achievers, and over achievers on the basis of the relationship between their intelligence-test scores and their general school achievement). (3) Data are collected that may *appear* to be interval or ratio in nature, but there is reason to believe that they do not contain equal units; hence, they are converted to the form of ranks (e.g., scores on a subjectively scored essay examination taken by a group of students undoubtedly do not represent equal units; each student is therefore given a rank in the group in lieu of his test score). In this chapter, we deal with both descriptive and inferential methods for treating data that can be represented as ranks. We begin with the case of a single rank-order variable available on the members of one group of subjects.   We then discuss summarization of such data in a rank-order classification table and an inferential goodness-of-fit test.   We turn next to bivariate data available on the members of one group and present a correlational approach to such data.   Finally, the case of two or more groups of subjects is covered and two hypothesis tests appropriate to this situation are developed.

8.1 *Univariate data for one group*

Assignment of ranks

Scores from rating scales and subjectively scored examinations do not typically produce equal-interval data.   For example, 15 essay-test scores are summarized in Table 8.1.   One score unit probably does not represent a constant amount of achievement, since the scores are subjectively determined by the teacher.

TABLE 8.1   Achievement-test scores for
N = 15 students

Student number	Score	Student number	Score	Student number	Score
1	80	6	86	11	81
2	86	7	75	12	80
3	77	8	84	13	71
4	93	9	89	14	86
5	76	10	63	15	75

Hence, it would be more defensible to convert scores of this type to ranks before carrying out further statistical analysis.   We would most likely be willing to consider the teacher's scores as a basis for ranking the students.   As a first step toward obtaining ranks, the original scores are placed in ascending or descending rank order.   Scores arranged in rank order are referred to as an *array*. Table 8.2 shows the array for the 15 test scores.   Ranks have been assigned to the scores, with rank 1 given to the numerically largest score.   (Of course, ranking could begin at either the high end or the low end of the array, but the largest score is traditionally given rank 1.)   When assigning ranks, the only difficulty is in dealing with tied scores.   The general rule is to assign the average of the positions at which the scores are tied to *each* tied score.   For example, three scores are tied at 86; these cover the positions 3, 4, and 5 in the distribution.   The average of 3, 4, and 5 is 4; hence, *each* score of 86 has been assigned

TABLE 8.2   Array for N = 15 test scores

Score	Rank	Score	Rank	Score	Rank
93	1	84	6	76	11
89	2	81	7	75	12½
86	4	80	8½	75	12½
86	4	80	8½	71	14
86	4	77	10	63	15

the rank of 4. Note that the next highest score (i.e., 84) receives the rank of 6, since ranks 3, 4, and 5 have already been used. Also, two scores of 80 appear and cover positions 8 and 9; the average rank, 8½, has been assigned to each score of 80 and the next highest score (i.e., 77) receives the rank of 10.

Rank-order classification tables

Assume that data have been generated in terms of rank-order categories; this may occur by direct assignment of subjects to preexisting categories (e.g., under achiever, average achiever, or over achiever) or by ranking subjects and then arbitrarily dividing the array into categories. The techniques presented in Chap. 2 for one-way classification tables could be utilized, and frequencies per category can be summarized for these ordinal data. For example, when a college instructor assigns letter grades to a class of students, the results can be summarized in a *rank-order classification table*. Table 8.3 illustrates such a table for a class of 58 students. Note that the rank-order classification table is simply a one-way classification table, but with ordered categories. As is shown in Table 8.3, the frequencies may be converted to proportions, as was

TABLE 8.3   Rank-order classification table for N = 58 beginning statistics students

Letter grade	Frequency	Proportion
A	5	.086
B	18	.310
C	21	.362
D	11	.190
F	3	.052
	58	1.000

previously discussed for univariate nominal data. Also, the modal category may be identified as the letter grade C. In addition to these procedures previously discussed, it is also possible to compute cumulative frequencies and cumulative proportions when dealing with ordinal data. A *cumulative frequency* is defined as the total *number* of cases *in or below* a stated category. Similarly. a *cumulative proportion* is the total *proportion* of cases *in or below* a stated category. Cumulative frequencies can be found systematically for all categories in a rank-order classification table by starting at the lowest category in the table (for the example, this is the letter grade F) and summing through the highest category, writing down subtotals as they are found. These cumulative frequencies can then be converted to cumulative proportions by dividing each cumulative frequency by N. To illustrate these techniques, Table 8.4 shows

TABLE 8.4   Cumulative frequency and cumulative proportion table for $N = 58$ beginning statistics students

Letter grade	Frequency	Cumulative frequency	Cumulative proportion
A	5	58	1.000
B	18	53	.914
C	21	35	.603
D	11	14	.241
F	3	3	.052
	58		

cumulative frequencies and cumulative proportions for the letter-grade data. The cumulative frequency of 35 for letter grade C is the sum of 21, 11, and 3. However, it could be found more simply by adding the frequency for letter grade C (that is, 21) to the previous cumulative frequency (that is, 14). Each cumulative frequency can be interpreted as the total number of students with letter grades *at or below* the given letter grade. Thus, a total of 14 students received letter grades of D or below, while 53 students received letter grades of B or below. The cumulative proportions were obtained by dividing each cumulative frequency by 58. Thus, $.603 = {}^{35}\!/_{58}$ is the proportion of students who received letter grades of C or below. In practice, it is easier to first find the decimal equivalent of $1/N$ and then *multiply* each cumulative frequency by this number. Thus, $^{1}\!/_{58} = .01724$ and $35(.01724) = .603$. Note that the cumulative frequency for the highest category must be N, whereas the cumulative proportion of this category must be 1.000.

Inferential procedures

When data from a random sample are classified in a rank-order classification table, the ordinality of the classification variable can be utilized in performing a goodness-of-fit test based on these data. The appropriate test is known as the *Kolmogorov-Smirnov one-sample test*. It requires that some assumption be made concerning the *cumulative proportions* for each category in the rank-order classification table. That is, the null hypothesis for a Kolmogorov-Smirnov test has the form $H_0: \psi_1 = a; \psi_2 = b; \ldots ; \psi_k = m$, where ψ_i is the hypothesized cumulative proportion through category k in the population, and a, b, $\ldots$, m are the specified numerical values for these cumulative proportions. A test statistic, d_{max}, is defined as the absolute value of the *largest* difference between a ψ value and the corresponding observed cumulative proportion. For relatively large sample sizes (n of 30 or more), critical values for a .05-significance-level test may be computed by means of the formula $C_{.95} = 1.36/\sqrt{n}$ and critical values at the .01 significance level from the formula $C_{.99} = 1.63/\sqrt{n}$. A table of critical values for smaller sample sizes is presented by Siegel (1956, Table E, page 251), but in most research situations the above formulas for approximate critical values will be satisfactory since relatively large sample sizes will be encountered. It should be noted that the test is nondirectional since the absolute value of the largest deviation between observed and expected cumulative proportion is utilized as the test statistic. If d_{max} exceeds the critical value at the chosen level of significance, the null hypothesis is rejected.

Let us illustrate this test for letter-grade data based on a random sample of 50 students drawn from the undergraduate enrollment in freshman English classes at a large college. Table 8.5 shows the frequencies and cumulative proportions of letter grades for these 50 students. If the distribution of letter grades follows the "curve" used by some instructors for alloting grades we would expect 7 percent A's, 24 percent B's, 38 percent C's, 24 percent D's, and

TABLE 8.5   Distribution of letter grades for 50 freshman English students

Letter grade	Observed frequency	Observed cumulative proportion
A	12	1.00
B	17	.76
C	12	.42
D	7	.18
F	2	.04
	50	

7 percent F's. These percentages are derived from considerations related to normal-curve theory, which was discussed in Chap. 5. If these percentages are converted to cumulative proportions, we can test the goodness of fit of the "curve" to the observed distribution of grades. The expected cumulative proportions are found by the same procedure used to find cumulative frequencies. We simply cumulate the "expected proportions" column. Thus, the expected cumulative proportion for letter grade C is .07 + .24 + .38, or .69. Table 8.6 shows this conversion from simple proportions to cumulative proportions and also presents d, the difference, in absolute value, between each pair of observed and expected cumulative proportions. Note that the null

TABLE 8.6   Determination of d values

Letter grade	Observed cumulative proportion	Expected proportion	Expected cumulative proportion	d
A	1.00	.07	1.00	.00
B	.76	.24	.93	.17
C	.42	.38	.69	.27
D	.18	.24	.31	.13
F	.04	.07	.07	.03

hypothesis can be written H_0: $\psi_1 = 1.00$, $\psi_2 = .93$, $\psi_3 = .69$, $\psi_4 = .31$, $\psi_5 = .07$. The maximum difference, in absolute value, between an observed and expected cumulative proportion is .27 for the C row. Thus, $d_{max} = .27$. At the .05 level of significance, the critical value is $C_{.95} = 1.36/\sqrt{50} = 1.36/7.0711 = .1923$. Since $d_{max} = .27$ exceeds .1923, the null hypothesis is rejected, and we conclude that this sample of 50 students does not represent a random sample from a population in which the distribution of letter grades follows the "curve." Inspection of Table 8.6 reveals that, in addition to the C row, relatively large discrepancies occur for the B and D letter grades.

8.2   Bivariate data

Descriptive procedures

In this section, we treat the case in which there are two quantified variables for each individual in a group; furthermore, both variables consist of rank-order data. It is quite common in practice for one of the sets of original scores to be on an interval or ratio scale but for the other set of scores to be ranks. It is then necessary to convert the first set of scores to ranks before further statistical analysis is carried out. Table 8.7 summarizes scores on a standardized achieve-

TABLE 8.7   *Raw scores and ranks on two variables*

Student number	Achievement-test raw score	Achievement-test rank	Class rank
1	24	8	9
2	20	10	8
3	46	4½	4
4	38	7	6
5	47	3	2½
6	21	9	10
7	50	1½	5
8	43	6	7
9	50	1½	1
10	46	4½	2½

ment test and class ranks for a group of students. To obtain ranks on both variables, it is necessary to convert the achievement-test scores to ranks. When ranking these scores, care should be taken to maintain the pairings of scores if the data are rearranged in any way.

The paired ranks in Table 8.7 constitute a *bivariate distribution of ranks.* As was true with categorical and measured data, it is possible to analyze rank-order data in order to determine the degree of association, or correlation, that exists between the two variables. When data are at least ordinal, a correlation may operate in either of two directions. (1) The *high* ranks on variable 1 tend to be associated with *high* ranks on variable 2, and the *low* ranks on variable 1 are associated with the *low* ranks on variable 2; or (2) the *high* ranks on variable 1 are associated with the *low* ranks on variable 2, and the *low* ranks on variable 1 are associated with the *high* ranks on variable 2. In the first case, we speak of the variables showing a direct, or positive, correlation; and in the second case, an inverse, or negative, correlation. Thus, the five pairs of ranks in Table 8.8 illustrate a perfect positive correlation, since the ranks are identical for any one person on the two variables. On the other hand, the data summarized in Table 8.9 show a perfect negative correlation between two variables, since a *high* rank on variable 1 is always associated with an equally *low* rank on variable 2, and vice versa. (Note that $R_2 = N - R_1 + 1 = 6 - R_1$.)

TABLE 8.8   A perfect positive correlation between two variables

Individual number	Ranks	
	R_1	R_2
1	3	3
2	2	2
3	1	1
4	5	5
5	4	4

TABLE 8.9   A perfect negative correlation between two variables

Individual number	Ranks	
	R_1	R_2
1	3	3
2	1	5
3	5	1
4	4	2
5	2	4

The 10 pairs of ranks presented in Table 8.7 show a positive correlation between the two tests; but this relationship is not perfect, since ranks are not identical on the two variables.

The degree of correlation between two rank-order variables may be expressed by a correlation coefficient. The most popular coefficient is known as the *rank-order correlation coefficient*.† It is derived from the Pearson product-moment correlation coefficient, which was presented in connection with measured data. In fact, the rank-order correlation coefficient is equivalent to the Pearsonian *r* obtained by using the ranks as if they were measured data. However, for computational purposes, a simpler approach is to use the following formula:

$$\rho_{\text{rank}} = 1 - \frac{6 \sum\limits_{i=1}^{N} d_i^2}{N^3 - N}$$

† This is also known as *Spearman's rank-difference correlation coefficient* and as *Spearman's rho.*

TABLE 8.10   *Examples of differing rank-order correlations*

Student number	Example 1			Example 2			Example 3			Example 4		
	R_1	R_2	d^2	R_1	R_2	d^2	R_1	R_2	d^2	R_1	R_2	d^2
1	1	1	0	1	4	9	1	5	16	1	2	1
2	2	2	0	2	1	1	2	4	4	2	3	1
3	3	3	0	3	3	0	3	3	0	3	1	4
4	4	4	0	4	5	1	4	2	4	4	5	1
5	5	5	0	5	2	9	5	1	16	5	4	1
$\sum_{i=1}^{5} d_i^2$			0			20			40			8
ρ_{rank}		$+1.00$			$.00$			-1.00			$+.60$	

where N is the number of individuals and d_i^2 is the squared *difference* between the pair of ranks for the ith individual. When $\rho_{rank} = 0$, this indicates that the two variables are uncorrelated, or independent of one another. If ρ_{rank} is positive, the variables show a positive relationship, with $\rho_{rank} = 1$ representing a perfect positive correlation. When ρ_{rank} is negative, an inverse or negative relationship exists, with $\rho_{rank} = -1$ representing a perfect negative correlation. To illustrate these situations, let us consider the examples in Table 8.10. For the first example,

$$\sum_{i=1}^{N} d_i^2 = 0 \quad \text{and} \quad \rho_{rank} = 1$$

for the second example,

$$\sum_{i=1}^{N} d_i^2 = 20 \quad \text{and} \quad \rho_{rank} = 1 - \frac{6(20)}{120} = 1 - 1 = 0$$

and for the third example,

$$\sum_{i=1}^{N} d_i^2 = 40 \quad \text{and} \quad \rho_{rank} = 1 - \frac{6(40)}{120} = 1 - 2 = -1$$

The fourth example shows an intermediate positive relationship with

$$\sum_{i=1}^{N} d_i^2 = 8 \quad \text{and} \quad \rho_{rank} = 1 - \frac{6(8)}{120} = 1 - .40 = .60$$

From the above examples, it can be seen that

$$\sum_{i=1}^{N} d_i^2$$

must be 0 in order for ρ_{rank} to be 1. That is, the pairs of ranks must be identical in order to obtain a perfect direct correlation. On the other hand, the maximum possible value for

$$\sum_{i=1}^{N} d_i{}^2$$

is $(N^3 - N)/3$. When this occurs, ρ_{rank} is -1. In order for ρ_{rank} to be 0,

$$\sum_{i=1}^{N} d_i{}^2$$

must equal $(N^3 - N)/6$.

Earlier in this textbook, when discussing the phi coefficient for categorical data and the Pearsonian coefficient for measured data, the notion that correlation was related to prediction was introduced. This idea can also be applied to the case in which the data are ranks. Consider first the case of two perfectly correlated variables. From example 1, Table 8.10, it is evident that the values of variable 2 can be predicted perfectly from variable 1. Thus, if you were told that an individual has rank 4 on variable 1, you could predict with certainty that his rank on variable 2 would also be 4. This is true because $R_1 = R_2$. Similarly, for two perfectly negatively correlated variables (example 2, Table 8.10), perfect prediction is possible because $R_2 = 6 - R_1$. Thus, for example, if $R_1 = 2$, then R_2 must be 4. For intermediate degrees of correlation, it is possible to build prediction equations, although this is rarely done. By following the logic developed in Chap. 7 for measured data, it is possible to derive estimates for the intercept and slope constants; the simplest approach is to utilize the equations $\hat{\alpha} = \mu_y - \beta\mu_x$ and $\hat{\beta} = \rho(\sigma_x/\sigma_y)$, where variable 1 is the X variable and variable 2 is the Y variable. We leave this derivation to the interested student. [Note that for any set of paired ranks, $\sigma_x = \sigma_y$ and $\mu_x = \mu_y = (N + 1)/2$.]

Inferential procedures

Computation of the rank-order correlation coefficient from a random sample of scores is accomplished in the same way as for a population of scores. The only change in the formula is that the population size N is replaced by the sample size n. If we adopt the symbol r_{rank} to represent a rank-order correlation coefficient computed from a sample, then

$$r_{\text{rank}} = 1 - \frac{6 \sum_{i=1}^{n} d_i{}^2}{n^3 - n}$$

Whenever a rank-order correlation coefficient is computed from a random sample of subjects, the magnitude of the coefficient is affected by sampling variability. That is, the value of r_{rank} from the sample cannot be expected to equal exactly the corresponding population value. When relatively small samples are involved, the effects of sampling variability are, of course, greater than for large samples.

In many practical situations, a researcher may be primarily interested in deciding simply if two variables are related to one another or if they are independent of one another in a specified population. Thus, the relevant null hypothesis would be $H_0: \rho_{rank} = 0$. The value of r_{rank} from a random sample of individuals from the specified population may be utilized to reach a decision concerning the tenability of the null hypothesis. For samples of size 10 or greater, an approximate test can be made by use of the t distributions presented in Table F.3. The relevant t statistic is

$$t = \frac{(r_{rank})(\sqrt{n-2})}{\sqrt{1 - r_{rank}^2}}$$

and the degrees of freedom are $n - 2$. For small sample sizes, critical values of r_{rank} are presented in Siegel (1956, Table P, page 284). Since the t statistic is the same as that for the Pearson correlation coefficient presented in Chap. 7, Table F.7 may be used to find critical values for r_{rank} also. Since ρ_{rank} may be either positive or negative, it is possible to construct directional tests by specifying whether ρ_{rank} is positive or negative. For example, if it were of interest only to decide whether ρ_{rank} was equal to zero or equal to some value *greater* than zero, a directional test would be used. The null hypothesis is still $H_0: \rho_{rank} = 0$. The alternative possibility of interest is $\rho_{rank} > 0$; that is, that ρ_{rank} is a positive, nonzero correlation.

As an example of the use of the t test, let us consider the data in Table 8.11. One variable consists of ranks based on the grade-point averages of 10 students randomly selected from one grade level of an elementary school. The second variable consists of ranks assigned to the students on a task related to creativity. The creativity ranks were assigned by a judge who was unaware of the students' academic standings. The purpose of the study was to determine if this particular creativity task was related to the previous achievement level of the students. The value of r_{rank} can be computed by formula as follows:

$$r_{rank} = 1 - \frac{6(63\frac{1}{2})}{10(99)} = 1 - \frac{381}{990} = 1 - .38 = .62$$

The relevant null hypothesis is $H_0: \rho_{rank} = 0$; and the test was planned as nondirectional, since it was possible that the two variables were either directly or inversely correlated. Inspection of Table F.7 for $df = 8$ reveals that a value of r_{rank} of at least .632 is required for rejection of the null hypothesis at the .05 level of significance. Thus, for this example, the null hypothesis must be retained at the .05 level. That is, *if* the null hypothesis is true, a value of r_{rank} as large as .62

TABLE 8.11   *Academic and creativity data for 10 students*

Student number	Academic ranks based on grade-point average	Creativity-test rank	d	d²
1	2	5	−3	9
2	4	1	3	9
3	8	10	−2	4
4	10	5	5	25
5	6½	9	−2½	6¼
6	6½	7	−½	¼
7	3	5	−2	4
8	5	3	2	4
9	1	2	−1	1
10	9	8	1	1

$$\sum_{i=1}^{10} d_i{}^2 = 63\tfrac{1}{2}$$

(or as small as −.62, since this is a nondirectional test) would occur more than 5 times in a hundred random samples of size 10.

8.3   Univariate data for two or more groups

Descriptive procedures

When two or more groups of subjects are compared on the basis of some variable, the researcher is generally interested in whether or not the groups differ from one another with respect to the values attained on the variable.   When entire populations are available for measurement purposes, decisions can be made in a straightforward manner.

Consider, first, the case in which data are arranged in ordinal categories for two or more groups of subjects.   If cumulative proportions are computed independently for each group of subjects, they may be compared directly to determine the relationships among the groups.   For example, Table 8.12 summarizes final-grade data for three professors each teaching different but comparable sections of one course during a semester.   Comparison of the "cumulative proportion" columns reveals some variation with respect to the distribution of F grades.   Professor 2 gave no F's, whereas Professor 1 gave 10 percent F's and Professor 3 gave 11 percent F's.   Continuing the comparison, it is evident that Professor 3 gave the largest proportion of D and below grades but that the proportions of B and below grades are very similar among the professors.

TABLE 8.12   *Final-grade data for three professors*

Grade	Professor 1	Cumulative proportion	Professor 2	Cumulative proportion	Professor 3	Cumulative proportion
A	3	1.00	2	1.00	2	1.00
B	10	.92	8	.93	8	.94
C	15	.68	12	.64	10	.72
D	8	.30	6	.21	12	.44
F	4	.10	0	.00	4	.11
	40		28		36	

The differing distributions of grades for the three professors may be construed in a number of ways. If the three classes of students were really comparable in learning potential at the beginning of the course and if the same or strictly comparable teaching procedures were used by the three professors, it seems reasonable to conclude that Professor 3 is the "hardest" grader. The "if's" in the previous sentence are highly speculative in any practical situation; and probably the strongest conclusion we would draw from these data is that the grade distributions did in fact differ among the three professors. Any conclusions concerning the reason for these discrepencies would necessarily be based on a good deal more information than has been presented here.

The above example illustrates the fact that comparisons of rank-order data for two or more groups may require relatively complex conclusions. That is, we often cannot conclude simply that one group is superior to some other group. Rather, the relationship among the groups may differ over the distribution.

In some research settings, the initial data collected by the researcher on two or more groups of subjects may appear to be interval or ratio in nature, but conversion to ranks may later become desirable. When this is the case, comparisons of the groups can be made by first ranking all scores, disregarding group membership, and then computing the sum of the ranks assigned within each group, or the average rank for each group. The degree to which these sums or averages differ is an indication of differences among the groups.

Inferential procedures

Two kinds of research data and the appropriate statistical tests for each are considered in this subsection. First, a technique is presented for testing the hypothesis that population cumulative proportions are identical for two populations when rank-order-classification data are available for two samples from these populations. Second, for data on two samples originally collected as interval or ratio scores, but converted to ranks for analysis, a technique will be presented for deciding whether or not these two samples were drawn from the same or comparable populations.

Let us first consider the case of data from two independent samples that have been reported in categorical form. If comparable categories are used, the *Kolmogorov-Smirnov two-sample test* can be used to test the hypothesis H_0: $\psi_{11} = \psi_{12}$; $\psi_{21} = \psi_{22}$; . . . ; $\psi_{k1} = \psi_{k2}$, where ψ_{ij} is the population cumulative proportion for category i and group j. As was true for the one-sample Kolmogorov-Smirnov test, the statistic of interest is a difference between cumulative proportions. For the present case, the statistic $d_{\max}$ is defined as the largest difference, in absolute value, between the cumulative proportions for the two samples. Critical values for relatively large samples (30 or more in each sample) can be found from the formulas for .05- and .01- significance-level tests:

$$C_{.95} = 1.36 \sqrt{\frac{1}{n_1} + \frac{1}{n_2}} \quad \text{and} \quad C_{.99} = 1.63 \sqrt{\frac{1}{n_1} + \frac{1}{n_2}}$$

The test is nondirectional since the absolute value of the difference is utilized as the test statistic. For smaller sample sizes, a table of critical values is presented in Siegel (1956, Table L, page 278). As an example of the application of the Kolmogorov-Smirnov two-sample test, let us consider the data summarized in Table 8.13. Sample 1 represents 30 randomly selected fresh-

TABLE 8.13   Rank-order classification data for two random samples of size 30

Letter grade	Sample 1 frequency	Sample 1 cumulative proportion	Sample 2 frequency	Sample 2 cumulative proportion	d
A	4	1.00	2	1.00	.00
B	8	.87	5	.93	.06
C	14	.60	11	.77	.17
D	2	.13	7	.40	.27
F	2	.07	5	.17	.10
	30		30		

man English grades assigned by professors at one college during one academic year; Sample 2 are grades assigned to freshman history students at the same college. The problem of interest is whether or not grade distributions for these two courses are the same. The null hypothesis can be written as

$$H: \psi_{A1} = \psi_{A2} \quad \psi_{B1} = \psi_{B2} \quad \psi_{C1} = \psi_{C2} \quad \psi_{D1} = \psi_{D2} \quad \psi_{F1} = \psi_{F2}$$

The test statistic is d_{max}, the largest absolute difference in cumulative proportion between the two samples, and this value is .27 from the D row of the table. For a nondirectional test at the .05 level of significance, the approximate critical value is $C_{.95} = 1.36 \sqrt{(\frac{1}{30}) + (\frac{1}{30})} = 1.36 \sqrt{.0667} = .35$. Since the computed value of d_{max} does not exceed this critical value, we must retain the null hypothesis and conclude that there is no evidence that freshman English and history grades are distributed differently at this college.

The second inferential procedure to be considered here, the Mann-Whitney U test, is applicable when the scores for two groups are initially considered to be on an interval or ratio scale of measurement but for various reasons the researcher wishes to analyze the scores as only ordinal data. Among the reasons that such a decision might be made are that (1) reconsideration of the measurement process might indicate that equal-interval assumptions are untenable but that scores do rank-order subjects; (2) a quick preliminary test might be desired before the more complex procedures for interval or ratio data are applied; or (3) certain assumptions required for measured-data tests might not be met, so that treatment as ordinal data might be undertaken as a "second-best" alternative.

The first step in the Mann-Whitney U test involves ranking together the scores for the two groups of subjects. That is, all scores are combined and ranks are assigned. Second, the ranks for the scores in each group are identified from the combined ranking, and statistics are computed from the sample sizes and the *sums* of ranks for the two groups. The derived statistics are

$$u_1 = n_1 n_2 + \frac{n_1(n_1 + 1)}{2} - R_1$$

$$u_2 = n_1 n_2 + \frac{n_2(n_2 + 1)}{2} - R_2$$

where n_1 and n_2 are the sample sizes for the two groups and R_1 and R_2 are the sums of ranks for the two groups. The null hypothesis states that the probability is .5 that a randomly selected score from population 1 will be larger than a randomly selected score from population 2. Thus, population 1 and 2 are of equal magnitudes in terms of score values.

For relatively large sample sizes (both n_1 and n_2 larger than 20), an approximate test statistic which is distributed as the unit normal distribution can be calculated. The appropriate statistic is $z = (u_1 - \mu_u)/\sigma_u$, where $\mu_u = n_1 n_2/2$ and $\sigma_u = \sqrt{n_1 n_2(n_1 + n_2 + 1)/12}$ and the value of z is referred to the table of the unit normal distribution (Table F.1) to determine the probability associated with the data (in the formula, if u_2 is used in place of u_1, the result will be of the opposite sign, but the magnitude of z will be the same). For smaller sample sizes, tables of critical values are presented by Siegel (1956, Tables J and K, pages 271–277).

To illustrate the use of the Mann-Whitney U test, let us consider the following example. A pilot study was designed to test the effects of two counseling techniques on the development of vocational goals among high school students. Independent raters interviewed the students before and after they were exposed to counseling and made judgments concerning the degree of improvement in level of vocational goals. The raters did not know which counseling techniques had been used with any individual student. Improvement scores for a total of 10 students, 6 in group 1 and 4 in group 2, are shown in Table 8.14. The ratings cannot reasonably be assumed to yield equal units

TABLE 8.14   Improvement scores for students exposed to two counseling techniques

Group 1	Group 2
18	18
21	13
14	9
23	5
8	
14	

of measurement; however, the rank order of improvement is probably accurately determined. Hence, inferential techniques for rank-order data are appropriate to this example. Also, the Mann-Whitney U test is proper, since we have data on two groups and they can be combined to obtain overall ranks.

In order to compute the u statistic, the scores from the two groups are combined and ranks are determined. Table 8.15 displays the results from this procedure. The sum of the ranks for the six scores in group 1 is found to be

$$R_1 = 1 + 2 + 3\tfrac{1}{2} + 5\tfrac{1}{2} + 5\tfrac{1}{2} + 9 = 26\tfrac{1}{2}$$

and for group 2, we have

$$R_2 = 3\tfrac{1}{2} + 7 + 8 + 10 = 28\tfrac{1}{2}$$

Then,

$$u_1 = n_1 n_2 + \frac{n_1 (n_1 + 1)}{2} - R_1 = 24 + \frac{6(7)}{2} - 26\tfrac{1}{2} = 18\tfrac{1}{2}$$

and

$$u_2 = n_1 n_2 + \frac{n_2(n_2 + 1)}{2} - R_2 = 24 + \frac{4(5)}{2} - 28\tfrac{1}{2} = 5\tfrac{1}{2}$$

TABLE 8.15   Combined ranking for 10 scores

Group	Score	Rank	Group	Score	Rank
1	23	1	1	14	$5\frac{1}{2}$
1	21	2	2	13	7
1	18	$3\frac{1}{2}$	2	9	8
2	18	$3\frac{1}{2}$	1	8	9
1	14	$5\frac{1}{2}$	2	5	10

The statement of the research problem implies that a nondirectional test is appropriate and we elect to use the .05 level of significance. Since the present sample sizes are less than 20, the tables in Siegel (1956) should be utilized for the test. These tables give critical values of 22 and 2, and since u_1 is less than 22 while u_2 is greater than 2, the null hypothesis must be retained. To illustrate the approximate z test, we can calculate $\mu_u = 6 \cdot 4/2 = 12$ and

$$\sigma_u = \sqrt{\frac{6 \cdot 4 \cdot 11}{12}} = \sqrt{22} = 4.6904$$

Thus, the test statistic is $z = (18.5 - 12)/4.6904 = 1.39$ which is nonsignificant using a nondirectional test at the .05 level of significance (since a value of z of at least 1.96 would be required to reject the null hypothesis).

The techniques presented in this chapter represent a limited sampling from those available for dealing with rank-order data.† In practice, these procedures are not widely used in educational research, and the great majority of research studies treat data as being measured. Thus, scores from an achievement test administered to students are typically interpreted as representing the number of correct responses, and this constitutes a ratio scale (that is, 20 items answered correctly represents twice as many correct responses as 10 items answered correctly). However, if the achievement-test scores are interpreted to represent the amount *learned* by the students, it is difficult to argue that the scores represent a ratio scale (e.g., has a student who scores 20 learned twice as much about the subject matter as a student who scores 10?). In fact, establishing that even equal intervals exist is usually impossible in such situations, and the data are, at best, ordinal with respect to the variable "amount learned." When data are derived from educational or psychological tests, the student should be careful to make interpretations that are consistent with the level of measurement assumed for the data. In chapters immediately preceding, techniques for treating data on interval or ratio scales of measurement were presented. The above cautions concerning the nature and inter-

† See Siegel [1956] for many additional procedures and references.

pretation of data should be kept in mind as these techniques are applied in practice.

Problems

8.1.   A teacher assigned the following scores to essay tests that she graded; assign the appropriate rank to each score.

36	28	42	28	20	24	36	28	40	20	40	18
24	20	28									

8.2.   The following grades were assigned by an instructor to his students: 6 A's; 14 B's; 9 C's; 4 D's; and 2 F's.   Construct a rank-order classification table and convert to proportions.

8.3.   Augment the table produced in Prob. 8.2 by including columns for cumulative frequency and cumulative proportion.

8.4.   A psychologist constructed a social-sensitivity scale for fourth graders; he used a large representative sample of fourth graders as a norm group and published the following fourth-grade norms:

Rating on social- sensitivity scale	Proportion
Very high	.10
High	.30
Average	.50
Low	.06
Very low	.04

A fourth-grade teacher administered the scale to her 35 students with the following results:

Very high	1
High	7
Average	14
Low	10
Very low	3

Is it likely that the differences between the social sensitivity of her group and the norm group are due to chance, or is her group significantly different from the norm group?

8.5.   The following pairs of scores were received by 10 students on 2 different tests.   Compute the rank-order correlation coefficient and test its significance using the .05 level of significance.

Student	Test 1	Test 2	Student	Test 1	Test 2
1	53	37	6	98	70
2	91	56	7	78	87
3	87	98	8	82	68
4	49	20	9	53	49
5	14	28	10	27	53

8.6.   Define positive correlation and negative correlation.

8.7.   How are scores ranked when two or more of the scores have equal values?

8.8.   For what reasons may a researcher decide to analyze data he "obtained on an interval or ratio scale" as if they were only ordinal data?

8.9.   Three vocational schools each used a different method for instructing students in arc welding.   A qualified welding expert rated the students in each of the schools upon completion of their training program.   A 5-point scale, with A representing superior ability and E representing very low ability, was used.   Utilizing the results presented below, compare the three methods of instruction by means of a descriptive procedure.

	Frequencies		
Rating	School 1	School 2	School 3
A	2	2	1
B	6	1	2
C	4	4	1
D	4	1	5
E	4	2	1

8.10.   Transform the following scores, which are on an interval scale, to ranks and compare the sum of ranks.   What procedure would you use if the groups were not the same size?

Group	
1	*2*
10	8
16	20
8	10
12	11
8	5
4	3

8.11. Twenty employees were randomly selected from each of two insurance companies. The probem of interest was whether or not the age distribution of employees in these two companies was the same. Using the Kolmogorov-Smirnov two-sample test, test this hypothesis at the .05 level of significance.

| Classification | Age | Frequency | |
		Company 1	Company 2
A	18–23	3	2
B	24–30	4	6
C	31–40	10	3
D	41–50	2	7
E	51–65	1	2

8.12. Assume that the data presented in Prob. 8.10 represent measures of job success. Using the Mann-Whitney U test, determine whether or not the special training program received by members of group 1 had an effect upon their success as employees. Employ the .05 level of significance.

technical appendix A

Units of Measurement and Significant Digits

Whenever approximate measurement is involved in a study, it is necessary to consider the precision with which scores are observed and recorded. It is obvious that the precision of the measurement process places restrictions on the precision of all future derived measures based on the scores resulting from measurement. Thus, if in a physical setting observations were taken to the nearest whole inch, it would be misleading (and incorrect) to express some resulting computation to, say, a thousandth of an inch.

Questions relating to the precision of measurement can be conveniently attacked by defining the *unit of measurement* underlying the recorded scores. For any given series of observations, the unit of measurement is the smallest division recorded by the measurement procedure. It is also the point in the measurement procedure at which rounding takes place. Thus, if a researcher were recording weights of a group of experimental subjects, he might utilize a scale that permits measurement to one-tenth of a pound. In effect, all observations would be rounded during the measurement process to the nearest $\frac{1}{10}$ pound. Thus, the unit of measurement for this series of observations would be one-tenth of a pound.

The simplest manner of determining the unit of measurement for a given set of scores is to become familiar with the measurement process that generated the scores, since the unit must be determined before observations can be taken. However, if an individual is confronted with a set of scores, it is usually possible to infer the unit of measurement from the way in which scores are reported (assuming that the scores have not been rounded to less precise form after observation; if additional rounding has taken place, one can only infer a *largest* unit of measurement).   For example, if scores such as 21.235, 45.198, etc., were reported, one could infer that the unit was .001 along the relevant scale. Of course, the researcher could have taken observations to the nearest .0001 or .00001 and have rounded them to thousandth prior to reporting them.

The number of *significant digits* in a measurement can be best understood by expressing observations in "scientific notation."   In this system, the observation is recorded entirely as a decimal quantity (usually between .1 and 1.0) with a suitable multiplication by a power of 10 to indicate the correct location of the decimal point.   That is, the general notation is $a \cdot 10^b$, where a is a decimal number and b is a positive or negative integer.   Consider the following examples.

Original number	Number in scientific notation
21.	$.21 \cdot 10^2$
21.4	$.214 \cdot 10^2$
1,063.57	$.106357 \cdot 10^4$
.684	$.684 \cdot 10^0$
.005	$.5 \cdot 10^{-2}$

In the first example, 21. represents a score recorded to a whole unit.   In scientific notation, the number is rewritten as .21 times a suitable power of 10. Since $(.21)(100) = 21.$, the necessary power of 10 is 2, since $10^2 = 100$.   Similarly, for the second and third examples, the decimal point is simply moved to the far left of the number, and the suitable power of 10 is the number of positions to the *left* that the decimal point is shifted (i.e., for the second example, the decimal point is shifted two positions to the left and the power of 10 is 2). In the fourth example, the original number is a completely decimal quantity and the decimal point is properly located; since $10^0 = 1$ by definition, we can use 10^0 in the scientific notation for the sake of a consistent system of notation.

The final example involves an original number that is entirely a decimal quantity but with zeros between the decimal point and the first relevant digit; in essence, the zeros are merely place holders. In scientific notation, the decimal quantity is rewritten as .5, and since the decimal point was moved two places to the *right*, the appropriate exponent is -2 (that is, $10^{-2} = \frac{1}{10^2} = \frac{1}{100} = .01$ and $(.5)(.01) = .005$, which is the original number).

Once a score has been rewritten in scientific notation, the number of significant digits in the score is simply the number of digits comprising the decimal quantity in the expression (ignoring the power of 10). Thus, the first example contains two significant digits since .21 contains two digits. The remaining examples contain three, six, three, and one significant digits, respectively. It should be noted that zeros that merely serve as place holders do not contribute to the number of significant digits. Thus, .005 contains only one significant digit. However, if the original number had been 1.005, it would contain four significant digits, since the number becomes $.1005 \cdot 10^1$ in scientific notation. Also, zeros can contribute to the number of significant digits if they occur at the end of the number.

In some cases, care must be taken in interpreting reported measurements, since the writer may adopt a convention of which the reader is unaware. For example, if I assert that the distance between two cities is 300 miles, it is unlikely that I mean that this figure is accurate to the nearest mile. Rather, the distance may be accurate to the nearest 10 or 20 miles; it is thus impossible to determine what convention I am utilizing without additional information.

Whenever approximate measurement is utilized, additional computations based on the observations should contain no more significant digits than the observation with the *least* number of significant digits. During intermediate computational steps, it is advisable to carry large numbers of digits in order to avoid the accumulation of rounding errors; but at a final stage, results should be rounded in conformity with this rule. Assume, as an example, that we wish to calculate the area of a triangular plot of land. The base of the triangular plot is measured and found to be 110.6 feet; the altitude is found to be 34.7 feet. Using the usual formula for the area of a triangle yields:

$$\text{Area} = \frac{1}{2}(110.6)(34.7) = \frac{1}{2}(3,837.82) = 1,918.91 \text{ square feet}$$

However, the base is measured to four significant digits (since $110.6 = .1106 \cdot 10^3$) whereas the altitude is measured to only three significant digits (since $34.7 = .347 \cdot 10^2$). Thus, the observation with the least number of significant digits is the altitude, and our result should be reported to only three significant digits. This could be given as 1,920 square feet; but this is ambiguous, since we can not tell if the final 0 is significant. A better system would be to report $.192 \cdot 10^4$ square feet as the area of the triangular plot. Note that the multiplicative factor, $\frac{1}{2}$, does not influence our determination of significant digits, since this factor is an exact number and not the result of approximate measurement.

In many educational applications, the observations are basically counts and represent exact, rather than approximate, measurement. Thus, if a student takes a multiple-choice examination and answers 32 out of 50 items correctly, the score of 32 is the exact number of correct responses and not the result of an approximate measuring process. If 10 students take the same examination, the average of their scores (found by summing the scores and dividing by the number of scores) can be legitimately expressed to several decimal places with little concern about significant digits. It is conventional in statistics to report only two or three decimal places for most derived quantities even when the observations are exact numbers resulting from counting procedures. It should be noted that there are occasions on which exact numbers resulting from such counting procedures as scoring objective examinations are treated nevertheless as if they represented points along a continuum (e.g., in constructing tables of the frequency distribution, frequency polygons, histograms, etc.).

technical appendix B

Goodness-of-fit Tests with Normal Distributions

Many of the inferential procedures dealing with measured data (i.e., data resulting from the use of interval or ratio scales of measurement) are based on an assumption of an underlying normal distribution in the population being studied. Also, it is often of interest to determine whether or not distributions of scores from standardized tests or from classroom tests given to large groups of students can be considered to represent normal distributions. In this appendix, we consider a test for the normality of a distribution of scores. The procedure is essentially a chi-square goodness-of-fit test, but the expected values have been computed to be in accordance with a hypothesized normal distribution. As we shall see, the degrees of freedom for this example present an interesting case.

The table on p. 224 summarizes test scores for a group of 172 students enrolled in an introductory quantitative methods course at the University of Maryland. The test yielding the scores was a 30-item unit examination; the frequency distribution has a class interval of 2 score units. The purpose of our analysis is to decide whether it is reasonable to consider these scores as representing a sample from a normal distribution. *Assuming* that the scores do come

Score limits	Frequency
24–25	5
22–23	11
20–21	29
18–19	31
16–17	37
14–15	33
12–13	16
10–11	7
8–9	3
Total	172

from a normally distributed population, it is possible to calculate expected frequencies for each of the score intervals (since all normal distributions have identical distributions in terms of percentages). Whereas the population mean and standard deviation are unknown, the values for the sample of 172 students are 16.99 and 3.46, respectively. These were computed from the raw scores, not from the grouped-data distribution. Taking, then, the normal distribution with a mean of 16.99 and standard deviation of 3.46, we can determine the expected frequency of scores in any given interval by reference to a table of values for the unit normal curve if we express the limits of the scores in z-score form. Thus, for example, the score limits for the first interval are 24 and 25. However, the true limits for scores falling in this interval are 23.5 and 25.5, since we must allow for rounding of scores. (This is also consistent with the fact that a normal distribution is always a continuous distribution.) Using the sample mean and standard deviation, we can express these limits in z-score form:

$$z_1 = \frac{(25.5 - 16.99)}{3.46} = 2.46$$

and

$$z_2 = \frac{23.5 - 16.99}{3.46} = 1.88$$

Turning to column $C(Z)$ of Table F.1, we see that the percentile rank for a z score of 2.46 is 99.31 and that for a z score of 1.88 is 96.99. Thus, 2.32 percent (or, 99.31 − 96.99) of the scores in a normal distribution would fall within the score interval with truth limits of 23.5 and 25.5. Since no score larger than 25 occurred in the distribution, we must also include in the highest interval that percentage under the normal distribution which was more extreme than anything actually observed. The percentage of cases beyond a z score of 2.46 is [from column $B(Z)$ of Table F.1] .69. Thus, we would expect 2.32 + .69, or 3.01, percent of the scores in or beyond our highest interval. The next score interval is 22 to 23 and has true limits of 21.5 and 23.5. The z scores corresponding to these limits are 1.30 and 1.88, respectively. Referring to column $C(Z)$ of Table F.1, we see that the percentile rank for a z score of 1.88 is 96.99 and for a z score of 1.30 is 90.32. Then, the percentage between these two points is 96.99 − 90.32, or 6.67. Continuing in this fashion for each interval yields the following results:

True limits	z scores for limits	Percentage in interval
23.5 and more	1.88 and more	3.01
21.5–23.5	1.30 and 1.88	6.67
19.5–21.5	.73 and 1.30	13.59
17.5–19.5	.15 and .73	20.77
15.5–17.5	−.43 and .15	22.60
13.5–15.5	−1.01 and −.43	17.74
11.5–13.5	−1.59 and −1.01	10.03
9.5–11.5	−2.16 and −1.59	4.05
9.5 and less	−2.16 and less	1.54

As was true for the highest interval, the lowest interval (7.5 to 9.5) also includes that part of the unit normal distribution more extreme than −2.74 z-score units.

These theoretical percentages for the intervals can now be used to compute expected frequencies (by multiplying by the sample size), and the usual chi-square goodness-of-fit statistic is easily calculated. The computational steps are shown below:

True limits	Frequency (f_i)	Theoretical percentage	Expected (f_i')	f_i^2/f_i'
23.5 and more	5	3.01	5.18	4.8263
21.5–23.5	11	6.67	11.47	10.5493
19.5–21.5	29	13.59	23.37	35.9863
17.5–19.5	31	20.77	35.72	26.9037
15.5–17.5	37	22.60	38.87	35.2200
13.5–15.5	33	17.74	30.51	35.6932
11.5–13.5	16	10.03	17.25	14.8406
9.5–11.5	7	4.05	6.97	7.0301
9.5 and less	3	1.54	2.65	3.3962
Total	172	100.00	171.99	174.4457

The sum of the final column, 174.4457, gives the first term in the computational formula for chi-square:

$$\chi^2 = \sum_{i=1}^{k} \frac{f_i^2}{f_i'} - n$$

Then $\chi^2 = 174.4457 - 172 = 2.45$.

We must now turn to a consideration of the appropriate degrees of freedom for this application of the chi-square goodness-of-fit test. As a general rule, if there are k categories of observations, then the degrees of freedom are $k - m$, where m is the number of different, independent restrictions placed on the expected values. One restriction always present is that the expected values sum to n, the sample size; that is,

$$\sum_{i=1}^{k} f_i' = n$$

In the present example, other restrictions are also implicit in the method for computing expected values. Since the population mean and standard deviation were unknown, we used the sample mean and standard deviation as estimates. Thus, our expected values are restricted to the case in which the mean is 16.99 and the standard deviation is 3.46, since these are the sample values used during computations. Therefore, we have two additional restrictions and $m = 3$ resulting in $k - 3$ degrees of freedom for the chi-square test. The three "missing" degrees of freedom can be conceptualized as representing one each for the total frequency of scores, the mean imposed on the normal distribution, and the standard deviation imposed on the normal distribution.

Since our data are grouped in nine categories, the degrees of freedom for the chi-square statistic are $9 - 3$, or 6. Entering Table F.4, we find that with 6 degrees of freedom, a value of at least 12.59 is required to reject the null hypothesis at the .05 level of significance; our value is obviously nonsignificant. It is of interest to be more specific concerning the null hypothesis for this example. In general terms, the null hypothesis can be stated by specifying the underlying normal distribution as having a mean of 16.99 and a standard deviation of 3.46; given this information, the expected values for any relevant frequency distribution could be determined by the methods developed above. In mathematical terms, we can write H_0: $\lambda_1 = 3.01$, $\lambda_2 = 6.67$, $\cdots$, $\lambda_9 = 1.54$, where λ_i is the theoretical percentage of cases in the ith category of the distribution.

technical appendix C

The t Test for the Two-sample Case

Although the one-dimensional analysis of variance is appropriate for measured data whenever there are two or more groups of experimental subjects, the student will frequently encounter references in the research literature to the use of t tests that can be applied when there are exactly two groups of subjects. In this appendix, we describe data situations for which the t test can be used. It should be emphasized at this point that the t test and the analysis of variance are equivalent to one another in the special case where there are two groups of experimental subjects.

Assume that an experimental design requires just two treatment groups—say, an experimental group and a control group. Further, suppose that an available group of n_t subjects is randomly divided between the two treatment conditions, so that n_1 subjects receive the experimental treatment and n_2 subjects receive the control condition (it is often desirable to divide the subject supply evenly between the two groups so that $n_1 = n_2 = n_t/2$). If we were interested in deciding whether or not the experimental treatment was superior to the control treatment in terms of some response measure, the relevant null hypothesis would be $\mu_1 = \mu_2$, and the analysis could be carried out by means of a one-dimensional analysis of variance.   The alternative analytical approach utilizing the t statistic is similar to the procedure described in Sec. 5.4, except that in the previous case the data represented resulted from matched pairs of experimental subjects (or the same experimental subject measured on two occasions); in the present case the data is based on independent groups of subjects.   Thus, instead of basing our test on the sampling distribution of a single sample mean, we must consider the sampling distribution of the *difference* between two sample means, $\bar{X}_1 - \bar{X}_2$.   For moderately large samples, this sampling distribution will be approximately normal and will center around the value $\mu_1 - \mu_2$, which is equal to 0, under our null hypothesis.   The standard deviation of the sampling distribution of $\bar{X}_1 - \bar{X}_2$ (that is, the standard error of the difference between the sample means) is, of course, generally unknown, since the population variances σ_1^2 and σ_2^2 have unknown values.   (If, however, these variances were known, the standard error would be

$$\sigma_{\bar{X}_1 - \bar{X}_2} = \sqrt{\frac{\sigma_1^2}{n_1} + \frac{\sigma_2^2}{n_2}}$$

and the null hypothesis could be tested by means of a z test of the form

$$z = \frac{\bar{X}_1 - \bar{X}_2}{\sigma_{\bar{X}_1 - \bar{X}_2}}$$

and tables of the unit normal curve would provide critical values).   As in previous examples, we can utilize the sample variances as estimates of the population variances and arrive at an appropriate estimate of the standard error of the difference.   As was true in the one-dimensional analysis of variance, this procedure depends upon the condition of homogeneity of variance; that is, $\sigma_1^2 = \sigma_2^2$ must hold.   This condition can be tested by means of the F test presented in Sec. 6.1.   Assuming that homogeneous variances characterize the data situation, the sample variances S_1^2 and S_2^2 can be pooled to arrive at a single estimate of $\sigma_{\bar{X}_1 - \bar{X}_2}$.   This estimate is

$$S_{\bar{X}_1 - \bar{X}_2} = S_e \sqrt{\frac{1}{n_1} + \frac{1}{n_2}}$$

where

$$S_e = \sqrt{\frac{(n_1 - 1)S_1^2 + (n_2 - 1)S_2^2}{n_1 + n_2 - 2}}$$

We are now in a position to characterize completely the sampling distribution of $\bar{X}_1 - \bar{X}_2$.   Under the null hypothesis, the arithmetic mean of the sampling distribution is 0 and (assuming the tenability of the assumption of homogeneity of variance) the estimated standard deviation is $S_{\bar{X}_1 - \bar{X}_2}$.   Thus, the observed sample mean difference, $\bar{X}_1 - \bar{X}_2$, can be located on the sampling distribution and an appropriate test set up.   By analogy with the one-sample t test (Sec. 5.4), the test statistic is simply $t = (\bar{X}_1 - \bar{X}_2)/S_{\bar{X}_1 - \bar{X}_2}$. Since the first sample is based on $n_1 - 1$ degrees of freedom and the second sample is based on $n_2 - 1$ degrees of freedom, the degrees of freedom for the t test are $(n_1 - 1) + (n_2 - 1)$, or $n_1 + n_2 - 2$. For a nondirectional test with $\alpha = .05$, the critical values would be the 2.5 and 97.5 percentiles of the t distribution with $n_1 + n_2 - 2$ degrees of freedom, and these values can be found in Table F.3. For the directional alternative hypothesis, $\mu_1 > \mu_2$, we would use the 95th percentile of the relevant t distribution, whereas the directional alternative hypothesis, $\mu_1 < \mu_2$, would require the use of the 5th percentile.

Consider the following data for an experimental and a control group:

Experimental group	Control group
25	19
20	22
31	14
27	16
19	11

Assume that a group of 10 subjects is randomly divided between the two treatment conditions and that the response measure represents some aspect of learning; also, assume that the researcher wishes to conduct a nondirectional test of the null hypothesis H_0: $\mu_1 = \mu_2$. We begin the analysis by testing the homogeneity-of-variance condition so that the pooling of sample variances can be legitimately carried out (although, with equal sample sizes, this condition is not critical to the validity of the test). The relevant summary statistics for the two groups are as shown in the table:

	Experimental group	Control group
Sum of Scores	122	82
Sum of Squared Scores	3076	1418
Arithmetic Mean	24.40	16.40
Variance	24.80	18.30

Using the sample variances, $S_1^2 = 24.80$ and $S_2^2 = 18.30$, the homogeneity-of-variance F statistic is simply $F = 24.80/18.30 = 1.36$, which with 5 and 5 degrees of freedom is nonsignificant at conventional levels. Thus, we conclude that there is no evidence to suspect that the condition $\sigma_1{}^2 = \sigma_2{}^2$ does not typify the populations with which we are working. Turning, then, to the t test, we see that the pooled variance is

$$\frac{(5-1)(24.80) + (5-1)(18.30)}{(5+5-2)} = \frac{172.40}{8} = 21.55$$

Thus, $S_e = \sqrt{21.55} = 4.64$ and

$$S_{\bar{x}_1-\bar{x}_2} = (4.64)\sqrt{(\tfrac{1}{5}) + (\tfrac{1}{5})} = 2.93$$

Finally, $t = (24.40 - 16.40)/2.93 = 8.00/2.93 = 2.73$. Selecting the .05 level of significance and entering the t table with 8 degrees of freedom, we find the critical values to be $+2.306$ and -2.306; we are able to reject the null hypothesis for these data and conclude that the experimental treatment is more effective than the control treatment.

We now consider the analysis of these same data by means of the analysis of variance; we shall demonstrate that the computed value of F from the analysis of variance is equal to the square of the t statistic (i.e., that $F = t^2$; this is a general relationship whenever the F statistic has 1 degree of freedom for its numerator). The relevant sums of squares are

$$SS_T = 4,494 - \frac{204^2}{10} = 4,494 - 4,161.60 = 332.40$$

$$SS_{\text{treat}} = \frac{122^2}{5} + \frac{82^2}{5} - \frac{204^2}{10} = 4,321.60 - 4,161.60 = 160.00$$

$$SS_{\text{error}} = SS_T - SS_{\text{treat}} = 332.40 - 160.00 = 172.40$$

Since there are two groups, the degrees of freedom associated with the treatment sum of square is 1; also, each group has five observations yielding 4 degrees of freedom per group for error. The analysis-of-variance summary table is:

Source	df	Sum of squares	Mean square	F
Treatment	1	160.00	160.00	7.42
Error	8	172.40	21.55	
Total	9	332.40		

The computed value of F is 7.42 and the square root of this quantity is

$$\sqrt{7.42} = 2.72$$

which is within a small rounding error of the t value computed earlier.

technical appendix D

z Tests for Proportions

In Chap. 2, we presented the one-sample chi-square test as an approximation to the exact binomial test and in Chap. 3, a chi-square test for two (or more) samples. The purpose of this appendix is to present alternate analytical approaches to two specific applications of these tests. Both applications involve situations in which the response dimension is dichotomized and the data are essentially binomial in form.

One-sample z test for proportions

A test based on z values associated with the unit normal distribution can be used in lieu of the chi-square approximation to the binomial test. The data situation is one in which a random sample of n observations is available and each observation is either a "success" or a "failure"; that is, the variable of interest is dichotomized. We assume that n_A of the observations are successes and the remaining $n_B = n - n_A$ observations are failures. Thus, the sample proportion of successes is $p_A = n_A/n$, and the proportion of failures is $p_B = n_B/n$ $= 1 - p_A$.

In terms of the general research situation, there is a relevant hypothesis concerning the true population proportion of successes; this value, under the null hypothesis, is π_A, and the true proportion of failures is $\pi_B = 1 - \pi_A$.

For moderately large numbers of observations, and if π_A is not too extreme (i.e., not too near to .00 or 1.00), the distribution of the number of successes is approximately normal. In fact, as n becomes large, the binomial distribution approaches a normal distribution (assuming π_A is a fixed value) as a limiting condition. This fact can be utilized to construct a test for the null hypothesis based on normal distributions. This test is based on the approach of specifying the parameters of the sampling distribution of p_A; this distribution, like the distribution of the number of successes in the parent population itself, is approximately normal for moderately large sample sizes. As intuititon would indicate, the arithmetic mean of the sampling distribution of p_A is equal to π_A, the hypothesized population proportion of successes. The standard deviation of this sampling distribution (i.e., the standard error of p_A) is equal to $\sqrt{\pi_A \pi_B / n}$. Thus, the observed proportion of successes in a sample of n observations can be located on this sampling distribution and its z-score equivalent computed; by reference to a table of the unit normal distribution, we can then determine either the directional or nondirectional probability associated with the sample outcome. The formula for computing the relevant z score is simply

$$z = \frac{p_A - \pi_A}{\sqrt{\pi_A \pi_B / n}}$$

As mentioned earlier in this appendix, the outcome based on a z statistic is equivalent to the chi-square approximation developed in Chap. 3. In fact, if both z and χ^2 were computed from the same data, the value of z would equal the square root of χ^2; that is, in this instance, $z = \sqrt{\chi^2}$.

By analogy with the one-sample chi-square test, the accuracy of the z test can be improved by introducing a correction term; this term is equivalent to reducing the difference between p_A and π_A by $.5/n$. An appropriate formula incorporating the correction (whose use is always recommended) is

$$z = \frac{|p_A - \pi_A| - .5/n}{\sqrt{\pi_A \pi_B / n}}$$

Note that the use of the absolute-value operator ensures that z will always be a positive quantity unless $|p_A - \pi_A|$ happens to be less than $.5/n$; in this event, the numerator of the expression should be set equal to 0 (and, of course, z becomes 0).

As an example of this z test, we return to the ESP experiment analyzed in Chap. 2 by both the exact binomial test and by use of the approximate chi-square statistic. In summary, the researcher observed 8 correct responses in a sample of 20 experimental subjects. The null hypothesis, based on a random guessing model, was $H_0: \pi_A = .25$. The value of p_A is $p_A = \frac{8}{20} = .40$, and the z statistic is

$$z = \frac{|.40 - .25| - .5/20}{\sqrt{(.25)(.75)/20}} \cdot = \frac{.15 - .025}{.0968} = \frac{.125}{.0968} = 1.29$$

Entering Table F.1 with $z = 1.29$, we find [from column $B(Z)$] that the directional probability associated with z is .0985 and the nondirectional probability [(from column $E(Z)$] is .1970. These figures are in agreement with that obtained from the use of chi-square; the computed chi-square value was 1.67 (using the corrected formula for chi-square) and the square root of chi-square is $\sqrt{1.67} = 1.29$, which is identical to our z value.

Two-sample z test for proportions

A similar z test is available in lieu of chi-square for the two-sample case when the response variable is a dichotomy (such as success and failure). In this case, we have one sample of n_1 observations of which n_{1A} are successes and a second sample of n_2 observations of which n_{2A} are successes; the respective proportions of successes are $p_1 = n_{1A}/n_1$ and $p_2 = n_{2A}/n_2$. The relevant null hypothesis in terms of corresponding population proportions is $H_0: \pi_1 = \pi_2$. The contrast of interest is the difference between p_1 and p_2; thus, our hypothesis-testing strategy must be to find a standard error for a test statistic involving the difference $p_1 - p_2$. Since π_1 and π_2 are unknown (by hypothesis they are equal but their numerical value is unspecified), we cannot compute a standard-error term in the same manner as for the one-sample test. However, information from the two samples can be pooled to define a suitable estimate. This estimate is $\sqrt{\bar{p}\bar{q}[(1/n_1) + (1/n_2)]}$, where $\bar{p} = (n_1 p_1 + n_2 p_2)/(n_1 + n_2)$ and $\bar{q} = 1 - \bar{p}$. Note that $\bar{p}$ is simply the overall proportion of successes in the two samples combined; thus, an alternate computing approach is

$$\bar{p} = \frac{n_{1A} + n_{2A}}{n_1 + n_2}$$

which is easily seen to be equivalent to the first definition. We can now define an appropriate z statistic:

$$z = \frac{p_1 - p_2}{\sqrt{\bar{p}\bar{q}[(1/n_1) + (1/n_2)]}}$$

As was true in the one-sample case (and with the corresponding chi-square tests), the accuracy of the z test is improved if a correction term is introduced.

The correction involves reducing the distance of both p_1 and p_2 from the pooled value $\bar{p}$. Assuming that p_1 is larger than p_2, the correction takes the form of changing p_1 to $p_1 - (.5/n_1)$ and changing p_2 to $p_2 + (.5/n_2)$. Since $\bar{p}$ must be numerically between p_1 and p_2, the correction has the effect of moving both p_1 and p_2 somewhat closer to $\bar{p}$ and thereby reducing the distance between them. The correction can be accomplished in one step by changing the numerator of the z formula to $|p_1 - p_2| - (n_1 + n_2)/2n_1n_2$.

As developed above, the z test is an alternative analysis to the chi-square test for a 2×2 contingency table (i.e., two groups of subjects responding on a dichotomized variable). The equivalence of the data situation can be seen from writing the contingency using the above notation.

	Response variable		
	Success, A	*Failure, B*	*Total*
Group 1	n_{1A}	n_{1B}	n_1
Group 2	n_{2A}	n_{2B}	n_2

If the z statistic and chi-square were both computed for one set of data, the value of z would be the square root of the value of chi-square. (In general, when chi-square has 1 degree of freedom, it is equivalent to the square of corresponding z values; however, since chi-square is a squared quantity and does not take direction into account, the 95th percentile of chi-square corresponds to the 97.5 percentile of z, the 99th percentile of chi-square corresponds to the 99.5 percentile of z, and so forth.) To demonstrate this relationship, let us consider the data below.

	Success	*Failure*	*Total*
Group 1	18	7	25
Group 2	11	14	25
Total	29	21	50

The sample proportions are $p_1 = 18/25 = .72$ and $p_2 = 11/25 = .44$; the overall proportion of success is $\bar{p} = 29/50 = .58$, and that of failure is

$$\bar{q} = 21/50 = .42 = 1 - .58$$

The pooled standard-error estimate is then

$$\sqrt{(.58)(.42)(.04 + .04)} = \sqrt{(.2436)(.08)} = \sqrt{.0195} = .1396$$

Finally,

$$z = \frac{|.72 - .44| - (50/1,250)}{.1396} = \frac{.24}{.1396} = 1.72$$

Entering Table F.1, we find that this z value is nonsignificant at the .05 level for a nondirectional test but significant at that level for a directional test with the alternative hypothesis $\pi_1 > \pi_2$.

The alternate analysis in terms of the chi-square statistic utilizes the simplified formula for 2×2 contingency tables:

$$\chi^2 = \frac{50(|252 - 77| - 25)^2}{(25)(25)(29)(21)}$$

$$= \frac{1,800}{609} = 2.96$$

The z value from the original analysis, when squared, is $(1.72)^2 = 2.96$, which is the same as the chi-square value.

technical appendix E

Relationships among Theoretical Distributions

In the course of presenting a variety of inferential statistical tests, reference has been made to several theoretical distributions. The most important of these are the unit normal (or z) distribution, the chi-square distribution, the "student" t distribution, and the F distribution. The purpose of this appendix is to summarize some mathematical relationships that exist among these distributions.

E. 1 Distribution of t and z

The distribution of the "student" t statistic with k degrees of freedom is

$$y = \frac{[(k-1)/2]!}{\sqrt{k\pi}\,[(k-2)/2]!}\,\frac{1}{[1 + (t^2/k)]^{(k+1)/2}}$$

As k, the degrees of freedom, grows large, this distribution converges to the unit normal distribution; thus, with infinite degrees of freedom, t is distributed as z. This relationship is apparent from inspection of a table of percentage points of the t distributions.

E.2   Distribution of chi-square and z

The distribution of chi-square with k degrees of freedom is (using u as the statistic)

$$Y = \frac{1}{[(k/2) - 1]!} \frac{1}{2^{k/2}} u^{(k/2)-1} e^{-\frac{1}{2}u}$$

In general, if the entering distributions of X's are normally distributed, chi-square represents the distribution of the *sum* of z^2 quantities. Thus, letting

$$u = \frac{\sum_{i=1}^{k} (X_i - \mu)^2}{\sigma^2}$$

we have a statistic distributed as chi-square with k degrees of freedom. Note that this representation of chi-square leads directly to the chi-square variance test presented in Sec. 5.5. If chi-square has $k = 1$ degrees of freedom, we have the distribution of a single z^2 value; since z itself is distributed as the unit normal distribution, this implies a direct relationship between chi-square and the unit normal curve. However, since it is z^2 that enters into chi-square, the relationship is such that the square root of the $1 - \alpha$ percentile of chi-square is equivalent to the $1 - \alpha/2$ percentile (and, by symmetry, to the $\alpha/2$ percentile if the negative square root is considered) of the unit normal curve (e.g., the 95th percentile of chi-square with 1 degree of freedom is 3.84, whereas the 97.5 percentile of the unit normal distribution is 1.96; note that $1.96^2 = 3.84$).

E.3   Distribution of F and t

Assume that two statistics, u and v, are each independently distributed as chi-square; that is, both u and v are defined as in Sec. E.2. Then, the ratio

$$\frac{u/k}{v/m}$$

is distributed as the F statistic with k degrees of freedom for the numerator and m degrees of freedom for the denominator. The distribution of F is given by

$$Y = \frac{[(k + m - 2)/2]!}{[(k - 2)/2]![(m - 2)/2]!} (k/m)^{k/2} \frac{F^{(k-2)/2}}{[1 + (kF/m)]^{(k+m)/2}}$$

When $k = 1$ (i.e., there is 1 degree of freedom for the numerator), the distribution of F is equivalent to the square of the corresponding t distribution. Thus, F with 1 and m degrees of freedom is equivalent to t^2 with m degrees of freedom (with, however, the $1 - \alpha$ percentile of F equivalent to the $1 - \alpha/2$, or the $\alpha/2$, percentile of t). For example, the 95th percentile of F with 1 and 20 degrees of freedom is 4.3513, whereas t with 20 degrees of freedom has its 97.5 percentile at 2.086 and its 2.5 percentile at -2.086; note that $2.086^2 = -2.086^2 = 4.3514$.

technical appendix F

TABLE F.1   Areas and ordinates of the unit normal distribution

Z	A(Z)	B(Z)	C(Z)	D(Z)	E(Z)	F(Z)
0.00	.0000	.5000	.5000	.0000	1.0000	.3989
0.01	.0040	.4960	.5040	.0080	.9920	.3989
0.02	.0080	.4920	.5080	.0160	.9840	.3989
0.03	.0120	.4880	.5120	.0240	.9760	.3988
0.04	.0160	.4840	.5160	.0320	.9680	.3986
0.05	.0199	.4801	.5199	.0398	.9602	.3984
0.06	.0239	.4761	.5239	.0478	.9522	.3982
0.07	.0279	.4721	.5279	.0558	.9442	.3980
0.08	.0319	.4681	.5319	.0638	.9362	.3977
0.09	.0359	.4641	.5359	.0718	.9282	.3973
0.10	.0398	.4602	.5398	.0796	.9204	.3970
0.11	.0438	.4562	.5438	.0876	.9124	.3965
0.12	.0478	.4522	.5478	.0956	.9044	.3961
0.13	.0517	.4483	.5517	.1034	.8966	.3956
0.14	.0557	.4443	.5557	.1114	.8886	.3951
0.15	.0596	.4404	.5596	.1192	.8808	.3945
0.16	.0636	.4364	.5636	.1272	.8728	.3939
0.17	.0675	.4325	.5675	.1350	.8650	.3932
0.18	.0714	.4286	.5714	.1428	.8572	.3925
0.19	.0753	.4247	.5753	.1506	.8494	.3918
0.20	.0793	.4207	.5793	.1586	.8414	.3910
0.21	.0832	.4168	.5832	.1664	.8336	.3902
0.22	.0871	.4129	.5871	.1742	.8258	.3894
0.23	.0910	.4090	.5910	.1820	.8180	.3885
0.24	.0948	.4052	.5948	.1896	.8104	.3876
0.25	.0987	.4013	.5987	.1974	.8026	.3867
0.26	.1026	.3974	.6026	.2052	.7948	.3857
0.27	.1064	.3936	.6064	.2128	.7872	.3847
0.28	.1103	.3897	.6103	.2206	.7794	.3836
0.29	.1141	.3859	.6141	.2282	.7718	.3825
0.30	.1179	.3821	.6179	.2358	.7642	.3814
0.31	.1217	.3783	.6217	.2434	.7566	.3802
0.32	.1255	.3745	.6255	.2510	.7490	.3790
0.33	.1293	.3707	.6293	.2586	.7414	.3778
0.34	.1331	.3669	.6331	.2662	.7338	.3765
0.35	.1368	.3632	.6368	.2736	.7264	.3752
0.36	.1406	.3594	.6406	.2812	.7188	.3739
0.37	.1443	.3557	.6443	.2886	.7114	.3725
0.38	.1480	.3520	.6480	.2960	.7040	.3712
0.39	.1517	.3483	.6517	.3034	.6966	.3697

TABLE F.1   (*continued*)

Z	A(Z)	B(Z)	C(Z)	C(Z)	E(Z)	F(Z)
0.40	.1554	.3446	.6554	.3108	.6892	.3683
0.41	.1591	.3409	.6591	.3182	.6818	.3668
0.42	.1628	.3372	.6628	.3256	.6744	.3653
0.43	.1664	.3336	.6664	.3328	.6672	.3637
0.44	.1700	.3300	.6700	.3400	.6600	.3621
0.45	.1736	.3264	.6736	.3472	.6528	.3605
0.46	.1772	.3228	.6772	.3544	.6456	.3589
0.47	.1808	.3192	.6808	.3616	.6384	.3572
0.48	.1844	.3156	.6844	.3688	.6312	.3555
0.49	.1879	.3121	.6879	.3758	.6242	.3538
0.50	.1915	.3085	.6915	.3830	.6170	.3521
0.51	.1950	.3050	.6950	.3900	.6100	.3503
0.52	.1985	.3015	.6985	.3970	.6030	.3485
0.53	.2019	.2981	.7019	.4038	.5962	.3467
0.54	.2054	.2946	.7054	.4108	.5892	.3448
0.55	.2088	.2912	.7088	.4176	.5824	.3429
0.56	.2123	.2877	.7123	.4246	.5754	.3410
0.57	.2157	.2843	.7157	.4314	.5686	.3391
0.58	.2190	.2810	.7190	.4380	.5620	.3372
0.59	.2224	.2776	.7224	.4448	.5552	.3352
0.60	.2257	.2743	.7257	.4514	.5486	.3332
0.61	.2291	.2709	.7291	.4582	.5418	.3312
0.62	.2324	.2676	.7324	.4648	.5352	.3292
0.63	.2357	.2643	.7357	.4714	.5286	.3271
0.64	.2389	.2611	.7389	.4778	.5222	.3251
0.65	.2422	.2578	.7422	.4844	.5156	.3230
0.66	.2454	.2546	.7454	.4908	.5092	.3209
0.67	.2486	.2514	.7486	.4972	.5028	.3187
0.68	.2517	.2483	.7517	.5034	.4966	.3166
0.69	.2549	.2451	.7549	.5098	.4902	.3144
0.70	.2580	.2420	.7580	.5160	.4840	.3123
0.71	.2611	.2389	.7611	.5222	.4778	.3101
0.72	.2642	.2358	.7642	.5284	.4716	.3079
0.73	.2673	.2327	.7673	.5346	.4654	.3056
0.74	.2704	.2296	.7704	.5408	.4592	.3034
0.75	.2734	.2266	.7734	.5468	.4532	.3011
0.76	.2764	.2236	.7764	.5528	.4472	.2989
0.77	.2794	.2206	.7794	.5588	.4412	.2966
0.78	.2823	.2177	.7823	.5646	.4354	.2943
0.79	.2852	.2148	.7852	.5704	.4296	.2920

TABLE F.1 (*continued*)

Z	A(Z)	B(Z)	C(Z)	C(Z)	E(Z)	F(Z)
0.80	.2881	.2119	.7881	.5762	.4238	.2897
0.81	.2910	.2090	.7910	.5820	.4180	.2874
0.82	.2939	.2061	.7939	.5878	.4122	.2850
0.83	.2967	.2033	.7967	.5934	.4066	.2827
0.84	.2995	.2005	.7995	.5990	.4010	.2803
0.85	.3023	.1977	.8023	.6046	.3954	.2780
0.86	.3051	.1949	.8051	.6102	.3898	.2756
0.87	.3078	.1922	.8078	.6156	.3844	.2732
0.88	.3106	.1894	.8106	.6212	.3788	.2709
0.89	.3133	.1867	.8133	.6266	.3734	.2685
0.90	.3159	.1841	.8159	.6318	.3682	.2661
0.91	.3186	.1814	.8186	.6372	.3628	.2637
0.92	.3212	.1788	.8212	.6424	.3576	.2613
0.93	.3238	.1762	.8238	.6476	.3524	.2589
0.94	.3264	.1736	.8264	.6528	.3472	.2565
0.95	.3289	.1711	.8289	.6578	.3422	.2541
0.96	.3315	.1685	.8315	.6630	.3370	.2516
0.97	.3340	.1660	.8340	.6680	.3320	.2492
0.98	.3365	.1635	.8365	.6730	.3270	.2468
0.99	.3389	.1611	.8389	.6778	.3222	.2444
1.00	.3413	.1587	.8413	.6826	.3174	.2420
1.01	.3438	.1562	.8438	.6876	.3124	.2396
1.02	.3461	.1539	.8461	.6922	.3078	.2371
1.03	.3485	.1515	.8485	.6970	.3030	.2347
1.04	.3508	.1492	.8508	.7016	.2984	.2323
1.05	.3531	.1469	.8531	.7062	.2938	.2299
1.06	.3554	.1446	.8554	.7108	.2892	.2275
1.07	.3577	.1423	.8577	.7154	.2846	.2251
1.08	.3599	.1401	.8599	.7198	.2802	.2227
1.09	.3621	.1379	.8621	.7242	.2758	.2203
1.10	.3643	.1357	.8643	.7286	.2714	.2179
1.11	.3665	.1335	.8665	.7330	.2670	.2155
1.12	.3686	.1314	.8686	.7372	.2628	.2131
1.13	.3708	.1292	.8708	.7416	.2584	.2107
1.14	.3729	.1271	.8729	.7458	.2542	.2083
1.15	.3749	.1251	.8749	.7498	.2502	.2059
1.16	.3770	.1230	.8770	.7540	.2460	.2036
1.17	.3790	.1210	.8790	.7580	.2420	.2012
1.18	.3810	.1190	.8810	.7620	.2380	.1989
1.19	.3830	.1170	.8830	.7660	.2340	.1965

TABLE F.1   (*continued*)

Z	A(Z)	B(Z)	C(Z)	D(Z)	E(Z)	F(Z)
1.20	.3849	.1151	.8849	.7698	.2302	.1942
1.21	.3869	.1131	.8869	.7738	.2262	.1919
1.22	.3888	.1112	.8888	.7776	.2224	.1895
1.23	.3907	.1093	.8907	.7814	.2186	.1872
1.24	.3925	.1075	.8925	.7850	.2150	.1849
1.25	.3944	.1056	.8944	.7888	.2112	.1826
1.26	.3962	.1038	.8962	.7924	.2076	.1804
1.27	.3980	.1020	.8980	.7960	.2040	.1781
1.28	.3997	.1003	.8997	.7994	.2006	.1758
1.29	.4015	.0985	.9015	.8030	.1970	.1736
1.30	.4032	.0968	.9032	.8064	.1936	.1714
1.31	.4049	.0951	.9049	.8098	.1902	.1691
1.32	.4066	.0934	.9066	.8132	.1868	.1669
1.33	.4082	.0918	.9082	.8164	.1836	.1647
1.34	.4099	.0901	.9099	.8198	.1802	.1626
1.35	.4115	.0885	.9115	.8230	.1770	.1604
1.36	.4131	.0869	.9131	.8262	.1738	.1582
1.37	.4147	.0853	.9147	.8294	.1706	.1561
1.38	.4162	.0838	.9162	.8324	.1676	.1539
1.39	.4177	.0823	.9177	.8354	.1646	.1518
1.40	.4192	.0808	.9192	.8384	.1616	.1497
1.41	.4207	.0793	.9207	.8414	.1586	.1476
1.42	.4222	.0778	.9222	.8444	.1556	.1456
1.43	.4236	.0764	.9236	.8472	.1528	.1435
1.44	.4251	.0749	.9251	.8502	.1498	.1415
1.45	.4265	.0735	.9265	.8530	.1470	.1394
1.46	.4279	.0721	.9279	.8558	.1442	.1374
1.47	.4292	.0708	.9292	.8584	.1416	.1354
1.48	.4306	.0694	.9306	.8612	.1388	.1334
1.49	.4319	.0681	.9319	.8638	.1362	.1315
1.50	.4332	.0668	.9332	.8664	.1336	.1295
1.51	.4345	.0655	.9345	.8690	.1310	.1276
1.52	.4357	.0643	.9357	.8714	.1286	.1257
1.53	.4370	.0630	.9370	.8740	.1260	.1238
1.54	.4382	.0618	.9382	.8764	.1236	.1219
1.55	.4394	.0606	.9394	.8788	.1212	.1200
1.56	.4406	.0594	.9406	.8812	.1188	.1182
1.57	.4418	.0582	.9418	.8836	.1164	.1163
1.58	.4429	.0571	.9429	.8858	.1142	.1145
1.59	.4441	.0559	.9441	.8882	.1118	.1127

TABLE F.1 *(continued)*

Z	A(Z)	B(Z)	C(Z)	D(Z)	E(Z)	F(Z)
1.60	.4452	.0548	.9452	.8904	.1096	.1109
1.61	.4463	.0537	.9463	.8926	.1074	.1092
1.62	.4474	.0526	.9474	.8948	.1052	.1074
1.63	.4484	.0516	.9484	.8968	.1032	.1057
1.64	.4495	.0505	.9495	.8990	.1010	.1040
1.65	.4505	.0495	.9505	.9010	.0990	.1023
1.66	.4515	.0485	.9515	.9030	.0970	.1006
1.67	.4525	.0475	.9525	.9050	.0950	.0989
1.68	.4535	.0465	.9535	.9070	.0930	.0973
1.69	.4545	.0455	.9545	.9090	.0910	.0957
1.70	.4554	.0446	.9554	.9108	.0892	.0940
1.71	.4564	.0436	.9564	.9128	.0872	.0925
1.72	.4573	.0427	.9573	.9146	.0854	.0909
1.73	.4582	.0418	.9582	.9164	.0836	.0893
1.74	.4591	.0409	.9591	.9182	.0818	.0878
1.75	.4599	.0401	.9599	.9198	.0802	.0863
1.76	.4608	.0392	.9608	.9216	.0784	.0848
1.77	.4616	.0384	.9616	.9232	.0768	.0833
1.78	.4625	.0375	.9625	.9250	.0750	.0818
1.79	.4633	.0367	.9633	.9266	.0734	.0804
1.80	.4641	.0359	.9641	.9282	.0718	.0790
1.81	.4649	.0351	.9649	.9298	.0702	.0775
1.82	.4656	.0344	.9656	.9312	.0688	.0761
1.83	.4664	.0336	.9664	.9328	.0672	.0748
1.84	.4671	.0329	.9671	.9342	.0658	.0734
1.85	.4678	.0322	.9678	.9356	.0644	.0721
1.86	.4686	.0314	.9686	.9372	.0628	.0707
1.87	.4693	.0307	.9693	.9386	.0614	.0694
1.88	.4699	.0301	.9699	.9398	.0602	.0681
1.89	.4706	.0294	.9706	.9412	.0588	.0669
1.90	.4713	.0287	.9713	.9426	.0574	.0656
1.91	.4719	.0281	.9719	.9438	.0562	.0644
1.92	.4726	.0274	.9726	.9452	.0548	.0632
1.93	.4732	.0268	.9732	.9464	.0536	.0620
1.94	.4738	.0262	.9738	.9476	.0524	.0608
1.95	.4744	.0256	.9744	.9488	.0512	.0596
1.96	.4750	.0250	.9750	.9500	.0500	.0584
1.97	.4756	.0244	.9756	.9512	.0488	.0573
1.98	.4761	.0239	.9761	.9522	.0478	.0562
1.99	.4767	.0233	.9767	.9534	.0466	.0551

TABLE F.1   (*continued*)

Z	A(Z)	B(Z)	C(Z)	D(Z)	E(Z)	F(Z)
2.00	.4772	.0228	.9772	.9544	.0456	.0540
2.01	.4778	.0222	.9778	.9556	.0444	.0529
2.02	.4783	.0217	.9783	.9566	.0434	.0519
2.03	.4788	.0212	.9788	.9576	.0424	.0508
2.04	.4793	.0207	.9793	.9586	.0414	.0498
2.05	.4798	.0202	.9798	.9596	.0404	.0488
2.06	.4803	.0197	.9803	.9606	.0394	.0478
2.07	.4808	.0192	.9808	.9616	.0384	.0468
2.08	.4812	.0188	.9812	.9624	.0376	.0459
2.09	.4817	.0183	.9817	.9634	.0366	.0449
2.10	.4821	.0179	.9821	.9642	.0358	.0440
2.11	.4826	.0174	.9826	.9652	.0348	.0431
2.12	.4830	.0170	.9830	.9660	.0340	.0422
2.13	.4834	.0166	.9834	.9668	.0332	.0413
2.14	.4838	.0162	.9838	.9676	.0324	.0404
2.15	.4842	.0158	.9842	.9684	.0316	.0396
2.16	.4846	.0154	.9846	.9692	.0308	.0387
2.17	.4850	.0150	.9850	.9700	.0300	.0379
2.18	.4854	.0146	.9854	.9708	.0292	.0371
2.19	.4857	.0143	.9857	.9714	.0286	.0363
2.20	.4861	.0139	.9861	.9722	.0278	.0355
2.21	.4864	.0136	.9864	.9728	.0272	.0347
2.22	.4868	.0132	.9868	.9736	.0264	.0339
2.23	.4871	.0129	.9871	.9742	.0258	.0332
2.24	.4875	.0125	.9875	.9750	.0250	.0325
2.25	.4878	.0122	.9878	.9756	.0244	.0317
2.26	.4881	.0119	.9881	.9762	.0238	.0310
2.27	.4884	.0116	.9884	.9768	.0232	.0303
2.28	.4887	.0113	.9887	.9774	.0226	.0297
2.29	.4890	.0110	.9890	.9780	.0220	.0290
2.30	.4893	.0107	.9893	.9786	.0214	.0283
2.31	.4896	.0104	.9896	.9792	.0208	.0277
2.32	.4898	.0102	.9898	.9796	.0204	.0270
2.33	.4901	.0099	.9901	.9802	.0198	.0264
2.34	.4904	.0096	.9904	.9808	.0192	.0258
2.35	.4906	.0094	.9906	.9812	.0188	.0252
2.36	.4909	.0091	.9909	.9818	.0182	.0246
2.37	.4911	.0089	.9911	.9822	.0178	.0241
2.38	.4913	.0087	.9913	.9826	.0174	.0235
2.39	.4916	.0084	.9916	.9832	.0168	.0229

TABLE F.1   (*continued*)

Z	A(Z)	B(Z)	C(Z)	D(Z)	E(Z)	F(Z)
2.40	.4918	.0082	.9918	.9836	.0164	.0224
2.41	.4920	.0080	.9920	.9840	.0160	.0219
2.42	.4922	.0078	.9922	.9844	.0156	.0213
2.43	.4925	.0075	.9925	.9850	.0150	.0208
2.44	.4927	.0073	.9927	.9854	.0146	.0203
2.45	.4929	.0071	.9929	.9858	.0142	.0198
2.46	.4931	.0069	.9931	.9862	.0138	.0194
2.47	.4932	.0068	.9932	.9864	.0136	.0189
2.48	.4934	.0066	.9934	.9868	.0132	.0184
2.49	.4936	.0064	.9936	.9872	.0128	.0180
2.50	.4938	.0062	.9938	.9876	.0124	.0175
2.51	.4940	.0060	.9940	.9880	.0120	.0171
2.52	.4941	.0059	.9941	.9882	.0118	.0167
2.53	.4943	.0057	.9943	.9886	.0114	.0163
2.54	.4945	.0055	.9945	.9890	.0110	.0158
2.55	.4946	.0054	.9946	.9892	.0108	.0154
2.56	.4948	.0052	.9948	.9896	.0104	.0151
2.57	.4949	.0051	.9949	.9898	.0102	.0147
2.58	.4951	.0049	.9951	.9902	.0098	.0143
2.59	.4952	.0048	.9952	.9904	.0096	.0139
2.60	.4953	.0047	.9953	.9906	.0094	.0136
2.61	.4955	.0045	.9955	.9910	.0090	.0132
2.62	.4956	.0044	.9956	.9912	.0088	.0129
2.63	.4957	.0043	.9957	.9914	.0086	.0126
2.64	.4959	.0041	.9959	.9918	.0082	.0122
2.65	.4960	.0040	.9960	.9920	.0080	.0119
2.66	.4961	.0039	.9961	.9922	.0078	.0116
2.67	.4962	.0038	.9962	.9924	.0076	.0113
2.68	.4963	.0037	.9963	.9926	.0074	.0110
2.69	.4964	.0036	.9964	.9928	.0072	.0107
2.70	.4965	.0035	.9965	.9930	.0070	.0104
2.71	.4966	.0034	.9966	.9932	.0068	.0101
2.72	.4967	.0033	.9967	.9934	.0066	.0099
2.73	.4968	.0032	.9968	.9936	.0064	.0096
2.74	.4969	.0031	.9969	.9938	.0062	.0093
2.75	.4970	.0030	.9970	.9940	.0060	.0091
2.76	.4971	.0029	.9971	.9942	.0058	.0088
2.77	.4972	.0028	.9972	.9944	.0056	.0086
2.78	.4973	.0027	.9973	.9946	.0054	.0084
2.79	.4974	.0026	.9974	.9948	.0052	.0081

TABLE F.1   (*continued*)

Z	A(Z)	B(Z)	C(Z)	D(Z)	E(Z)	F(Z)
2.80	.4974	.0026	.9974	.9948	.0052	.0079
2.81	.4975	.0025	.9975	.9950	.0050	.0077
2.82	.4976	.0024	.9976	.9952	.0048	.0075
2.83	.4977	.0023	.9977	.9954	.0046	.0073
2.84	.4977	.0023	.9977	.9954	.0046	.0071
2.85	.4978	.0022	.9978	.9956	.0044	.0069
2.86	.4979	.0021	.9979	.9958	.0042	.0067
2.87	.4979	.0021	.9979	.9958	.0042	.0065
2.88	.4980	.0020	.9980	.9960	.0040	.0063
2.89	.4981	.0019	.9981	.9962	.0038	.0061
2.90	.4981	.0019	.9981	.9962	.0038	.0060
2.91	.4982	.0018	.9982	.9964	.0036	.0058
2.92	.4982	.0018	.9982	.9964	.0036	.0056
2.93	.4983	.0017	.9983	.9966	.0034	.0055
2.94	.4984	.0016	.9984	.9968	.0032	.0053
2.95	.4984	.0016	.9984	.9968	.0032	.0051
2.96	.4985	.0015	.9985	.9970	.0030	.0050
2.97	.4985	.0015	.9985	.9970	.0030	.0048
2.98	.4986	.0014	.9986	.9972	.0028	.0047
2.99	.4986	.0014	.9986	.9972	.0028	.0046
3.00	.4987	.0013	.9987	.9974	.0026	.0044
3.01	.4987	.0013	.9987	.9974	.0026	.0043
3.02	.4987	.0013	.9987	.9974	.0026	.0042
3.03	.4988	.0012	.9988	.9976	.0024	.0040
3.04	.4988	.0012	.9988	.9976	.0024	.0039
3.05	.4989	.0011	.9989	.9978	.0022	.0038
3.06	.4989	.0011	.9989	.9978	.0022	.0037
3.07	.4989	.0011	.9989	.9978	.0022	.0036
3.08	.4990	.0010	.9990	.9980	.0020	.0035
3.09	.4990	.0010	.9990	.9980	.0020	.0034
3.10	.4990	.0010	.9990	.9980	.0020	.0033
3.11	.4991	.0009	.9991	.9982	.0018	.0032
3.12	.4991	.0009	.9991	.9982	.0018	.0031
3.13	.4991	.0009	.9991	.9982	.0018	.0030
3.14	.4992	.0008	.9992	.9984	.0016	.0029
3.15	.4992	.0008	.9992	.9984	.0016	.0028
3.16	.4992	.0008	.9992	.9984	.0016	.0027
3.17	.4992	.0008	.9992	.9984	.0016	.0026
3.18	.4993	.0007	.9993	.9986	.0014	.0025
3.19	.4993	.0007	.9993	.9986	.0014	.0025
3.20	.4993	.0007	.9993	.9986	.0014	.0024

TABLE F.2   *Percentiles of the unit normal distribution*

P	Z	F(Z)	P	Z	F(Z)
.005	-2.5758	.0145	.205	-0.8239	.2841
.010	-2.3263	.0267	.210	-0.8064	.2882
.015	-2.1701	.0379	.215	-0.7892	.2922
.020	-2.0537	.0484	.220	-0.7722	.2961
.025	-1.9600	.0584	.225	-0.7554	.2999
.030	-1.8808	.0680	.230	-0.7388	.3036
.035	-1.8119	.0773	.235	-0.7225	.3073
.040	-1.7507	.0862	.240	-0.7063	.3109
.045	-1.6954	.0948	.245	-0.6903	.3144
.050	-1.6449	.1031	.250	-0.6745	.3178
.055	-1.5982	.1112	.255	-0.6588	.3211
.060	-1.5548	.1191	.260	-0.6433	.3244
.065	-1.5141	.1268	.265	-0.6280	.3275
.070	-1.4758	.1343	.270	-0.6128	.3306
.075	-1.4395	.1416	.275	-0.5978	.3337
.080	-1.4051	.1487	.280	-0.5828	.3366
.085	-1.3722	.1556	.285	-0.5681	.3395
.090	-1.3408	.1624	.290	-0.5534	.3423
.095	-1.3106	.1690	.295	-0.5388	.3450
.100	-1.2816	.1755	.300	-0.5244	.3477
.105	-1.2536	.1818	.305	-0.5101	.3503
.110	-1.2265	.1880	.310	-0.4959	.3528
.115	-1.2004	.1941	.315	-0.4817	.3552
.120	-1.1750	.2000	.320	-0.4677	.3576
.125	-1.1503	.2059	.325	-0.4538	.3599
.130	-1.1264	.2115	.330	-0.4399	.3621
.135	-1.1031	.2171	.335	-0.4261	.3643
.140	-1.0803	.2226	.340	-0.4125	.3664
.145	-1.0581	.2279	.345	-0.3989	.3684
.150	-1.0364	.2332	.350	-0.3853	.3704
.155	-1.0152	.2383	.355	-0.3719	.3723
.160	-0.9945	.2433	.360	-0.3585	.3741
.165	-0.9741	.2482	.365	-0.3451	.3759
.170	-0.9542	.2531	.370	-0.3319	.3776
.175	-0.9346	.2578	.375	-0.3186	.3792
.180	-0.9154	.2624	.380	-0.3055	.3808
.185	-0.8965	.2669	.385	-0.2924	.3822
.190	-0.8779	.2714	.390	-0.2793	.3837
.195	-0.8596	.2757	.395	-0.2663	.3850
.200	-0.8416	.2800	.400	-0.2533	.3863

TABLE F.2   (*continued*)

P	Z	F(Z)	P	Z	F(Z)
.405	−0.2404	.3876	.605	0.2663	.3850
.410	−0.2275	.3887	.610	0.2793	.3837
.415	−0.2147	.3899	.615	0.2924	.3822
.420	−0.2019	.3909	.620	0.3055	.3808
.425	−0.1891	.3919	.625	0.3186	.3792
.430	−0.1764	.3928	.630	0.3319	.3776
.435	−0.1637	.3936	.635	0.3451	.3759
.440	−0.1510	.3944	.640	0.3585	.3741
.445	−0.1383	.3951	.645	0.3719	.3723
.450	−0.1257	.3958	.650	0.3853	.3704
.455	−0.1130	.3964	.655	0.3989	.3684
.460	−0.1004	.3969	.660	0.4125	.3664
.465	−0.0878	.3974	.665	0.4261	.3643
.470	−0.0753	.3978	.670	0.4399	.3621
.475	−0.0627	.3982	.675	0.4538	.3599
.480	−0.0502	.3984	.680	0.4677	.3576
.485	−0.0376	.3987	.685	0.4817	.3552
.490	−0.0251	.3988	.690	0.4959	.3528
.495	−0.0125	.3989	.695	0.5101	.3503
.500	0.0000	.3989	.700	0.5244	.3477
.505	0.0125	.3989	.705	0.5388	.3450
.510	0.0251	.3988	.710	0.5534	.3423
.515	0.0376	.3987	.715	0.5681	.3395
.520	0.0502	.3984	.720	0.5828	.3366
.525	0.0627	.3982	.725	0.5978	.3337
.530	0.0753	.3978	.730	0.6128	.3306
.535	0.0878	.3974	.735	0.6280	.3275
.540	0.1004	.3969	.740	0.6433	.3244
.545	0.1130	.3964	.745	0.6588	.3211
.550	0.1257	.3958	.750	0.6745	.3178
.555	0.1383	.3951	.755	0.6903	.3144
.560	0.1510	.3944	.760	0.7063	.3109
.565	0.1637	.3936	.765	0.7225	.3073
.570	0.1764	.3928	.770	0.7388	.3036
.575	0.1891	.3919	.775	0.7554	.2999
.580	0.2019	.3909	.780	0.7722	.2961
.585	0.2147	.3899	.785	0.7892	.2922
.590	0.2275	.3887	.790	0.8064	.2882
.595	0.2404	.3876	.795	0.8239	.2841
.600	0.2533	.3863	.800	0.8416	.2800

TABLE F.2   (*continued*)

P	Z	F(Z)
.805	0.8596	.2757
.810	0.8779	.2714
.815	0.8965	.2669
.820	0.9154	.2624
.825	0.9346	.2578
.830	0.9542	.2531
.835	0.9741	.2482
.840	0.9945	.2433
.845	1.0152	.2383
.850	1.0364	.2332
.855	1.0581	.2279
.860	1.0803	.2226
.865	1.1031	.2171
.870	1.1264	.2115
.875	1.1503	.2059
.880	1.1750	.2000
.885	1.2004	.1941
.890	1.2265	.1880
.895	1.2536	.1818
.900	1.2816	.1755
.905	1.3106	.1690
.910	1.3408	.1624
.915	1.3722	.1556
.920	1.4051	.1487
.925	1.4395	.1416
.930	1.4758	.1343
.935	1.5141	.1268
.940	1.5548	.1191
.945	1.5982	.1112
.950	1.6449	.1031
.955	1.6954	.0948
.960	1.7507	.0862
.965	1.8119	.0773
.970	1.8808	.0680
.975	1.9600	.0584
.980	2.0537	.0484
.985	2.1701	.0379
.990	2.3263	.0267
.995	2.5758	.0145

TABLE F.3   *Selected percentiles of t distributions*

PERCENTILE

DF	.005	.01	.025	.05	.95	.975	.99	.995
1	-63.657	-31.821	-12.706	-6.314	6.314	12.706	31.821	63.657
2	-9.925	-6.965	-4.303	-2.920	2.920	4.303	6.965	9.925
3	-5.841	-4.541	-3.182	-2.353	2.353	3.182	4.541	5.841
4	-4.604	-3.747	-2.776	-2.132	2.132	2.776	3.747	4.604
5	-4.032	-3.365	-2.571	-2.015	2.015	2.571	3.365	4.032
6	-3.707	-3.143	-2.447	-1.943	1.943	2.447	3.143	3.707
7	-3.499	-2.998	-2.365	-1.895	1.895	2.365	2.998	3.499
8	-3.355	-2.896	-2.306	-1.860	1.860	2.306	2.896	3.499
9	-3.250	-2.821	-2.262	-1.833	1.833	2.262	2.821	3.355
10	-3.169	-2.764	-2.228	-1.812	1.812	2.228	2.764	3.250
11	-3.106	-2.718	-2.201	-1.796	1.796	2.201	2.718	3.169
12	-3.055	-2.681	-2.179	-1.782	1.782	2.179	2.681	3.106
13	-3.012	-2.650	-2.160	-1.771	1.771	2.160	2.650	3.055
14	-2.977	-2.624	-2.145	-1.761	1.761	2.145	2.624	3.012
15	-2.947	-2.602	-2.131	-1.753	1.753	2.131	2.602	2.977
16	-2.921	-2.583	-2.120	-1.746	1.746	2.120	2.583	2.947
17	-2.898	-2.567	-2.110	-1.740	1.740	2.110	2.567	2.921
18	-2.878	-2.552	-2.101	-1.734	1.734	2.101	2.552	2.898
19	-2.861	-2.539	-2.093	-1.729	1.729	2.093	2.539	2.878
20	-2.845	-2.528	-2.086	-1.725	1.725	2.086	2.528	2.861
21	-2.831	-2.518	-2.080	-1.721	1.721	2.080	2.518	2.831
22	-2.819	-2.508	-2.074	-1.717	1.717	2.074	2.508	2.819
23	-2.807	-2.500	-2.069	-1.714	1.714	2.069	2.500	2.807
24	-2.797	-2.492	-2.064	-1.711	1.711	2.064	2.492	2.797
25	-2.787	-2.485	-2.060	-1.708	1.708	2.060	2.485	2.787
26	-2.779	-2.479	-2.056	-1.706	1.706	2.056	2.479	2.779
27	-2.771	-2.473	-2.052	-1.703	1.703	2.052	2.473	2.771
28	-2.763	-2.467	-2.048	-1.701	1.701	2.048	2.467	2.763
29	-2.756	-2.462	-2.045	-1.699	1.699	2.045	2.462	2.756
30	-2.750	-2.457	-2.042	-1.697	1.697	2.042	2.457	2.750
40	-2.704	-2.423	-2.021	-1.684	1.684	2.021	2.423	2.704
50	-2.678	-2.403	-2.009	-1.676	1.676	2.009	2.403	2.678
60	-2.660	-2.390	-2.000	-1.671	1.671	2.000	2.390	2.660
80	-2.639	-2.374	-1.990	-1.664	1.664	1.990	2.374	2.639
100	-2.626	-2.365	-1.984	-1.660	1.660	1.984	2.365	2.626
120	-2.617	-2.358	-1.980	-1.658	1.658	1.980	2.358	2.617
200	-2.601	-2.345	-1.972	-1.653	1.653	1.972	2.345	2.601
500	-2.586	-2.334	-1.965	-1.648	1.648	1.965	2.334	2.601
∞	-2.576	-2.326	-1.960	-1.645	1.645	1.960	2.326	2.576

TABLE F.4   Selected percentiles of Chi-square distributions

PERCENTILE

DF	.005	.01	.025	.05	.95	.975	.99	.995
1	.00004	.00016	.00098	.00393	3.84	5.02	6.63	7.88
2	.0100	.0201	.0506	.103	5.99	7.38	9.21	10.60
3	.072	.115	.216	.352	7.81	9.35	11.34	12.84
4	.207	.297	.484	.711	9.49	11.14	13.28	14.86
5	.412	.554	.831	1.145	11.07	12.83	15.09	16.75
6	.676	.872	1.24	1.64	12.59	14.45	16.81	18.55
7	.989	1.24	1.69	2.17	14.07	16.01	18.48	20.28
8	1.34	1.65	2.18	2.73	15.51	17.53	20.09	21.96
9	1.73	2.09	2.70	3.33	16.92	19.02	21.67	23.59
10	2.16	2.56	3.25	3.94	18.31	20.48	23.21	25.19
11	2.60	3.05	3.82	4.57	19.68	21.92	24.72	26.76
12	3.07	3.57	4.40	5.23	21.03	23.34	26.22	28.30
13	3.57	4.11	5.01	5.89	22.36	24.74	27.69	29.82
14	4.07	4.66	5.63	6.57	23.68	26.12	29.14	31.32
15	4.60	5.23	6.26	7.26	25.00	27.49	30.58	32.80
16	5.14	5.81	6.91	7.96	26.30	28.85	32.00	34.27
17	5.70	6.41	7.56	8.67	27.59	30.19	33.41	35.72
18	6.26	7.01	8.23	9.39	28.87	31.53	34.81	37.16
19	6.84	7.63	8.91	10.12	30.14	32.85	36.19	38.58
20	7.43	8.26	9.59	10.85	31.41	34.17	37.57	40.00
21	8.03	8.90	10.28	11.59	32.67	35.48	38.93	41.40
22	8.64	9.54	10.98	12.34	33.92	36.78	40.29	42.80
23	9.26	10.20	11.69	13.09	35.17	38.08	41.64	44.18
24	9.89	10.86	12.40	13.85	36.42	39.36	42.98	45.56
25	10.52	11.52	13.12	14.61	37.65	40.65	44.31	46.93
26	11.16	12.20	13.84	15.38	38.89	41.92	45.64	48.29
27	11.81	12.88	14.57	16.15	40.11	43.19	46.96	49.64
28	12.46	13.56	15.31	16.93	41.34	44.46	48.28	50.99
29	13.21	14.26	16.05	17.71	42.56	45.72	49.59	52.34
30	13.79	14.95	16.79	18.49	43.77	46.98	50.89	53.67
40	20.71	22.16	24.43	26.51	55.76	59.34	63.69	66.77
50	27.99	29.71	32.36	34.76	67.50	71.42	76.15	79.49
60	35.53	37.48	40.48	43.19	79.08	83.30	88.38	91.95
70	43.28	45.44	48.76	51.74	90.53	95.02	100.4	104.2
80	51.17	53.54	57.15	60.39	101.9	106.6	112.3	116.3
90	59.20	61.75	65.65	69.13	113.1	118.1	124.1	128.3
100	67.33	70.06	74.22	77.93	124.3	129.6	135.8	140.2
120	83.85	86.92	91.58	95.70	146.6	152.2	159.0	163.6

TABLE F.5   *Selected percentiles of F distribution*

DF						PERCENTILE			
NUM	DEN	.005	.01	.025	.05	.95	.975	.99	.995
1	1	.0001	.0002	.0015	.0062	161.45	647.79	4052.2	16211.
	2	.0000	.0002	.0012	.0050	18.513	38.506	98.503	198.50
	3	.0000	.0002	.0012	.0046	10.128	17.443	34.116	55.552
	4	.0000	.0002	.0011	.0044	7.7086	12.218	21.198	31.333
	5	.0000	.0002	.0011	.0043	6.6079	10.007	16.258	22.785
	6	.0000	.0002	.0011	.0043	5.9874	8.8131	13.745	18.635
	7	.0000	.0002	.0010	.0042	5.5914	8.0727	12.246	16.236
	8	.0000	.0002	.0010	.0042	5.3177	7.5709	11.259	14.688
	9	.0000	.0002	.0010	.0042	5.1174	7.2093	10.561	13.614
	10	.0000	.0002	.0010	.0041	4.9646	6.9367	10.044	12.826
	12	.0000	.0002	.0010	.0041	4.7472	6.5538	9.3302	11.754
	15	.0000	.0002	.0010	.0041	4.5431	6.1995	8.6831	10.798
	20	.0000	.0002	.0010	.0040	4.3513	5.8715	8.0960	9.9439
	24	.0000	.0002	.0010	.0040	4.2597	5.7167	7.8229	9.5513
	30	.0000	.0002	.0010	.0040	4.1709	5.5675	7.5625	9.1797
	40	.0000	.0002	.0010	.0040	4.0848	5.4239	7.3141	8.8278
	60	.0000	.0002	.0010	.0040	4.0012	5.2857	7.0771	8.4946
	120	.0000	.0002	.0010	.0039	3.9201	5.1524	6.8510	8.1790
	∞	.0000	.0002	.0010	.0039	3.8415	5.0239	6.6349	7.8794
2	1	.0050	.0101	.0260	.0540	199.50	799.50	4999.5	20000.
	2	.0050	.0101	.0256	.0526	19.000	39.000	99.000	199.00
	3	.0050	.0101	.0255	.0522	9.5521	16.044	30.817	49.799
	4	.0050	.0101	.0255	.0520	6.9443	10.649	18.000	26.284
	5	.0050	.0101	.0254	.0518	5.7861	8.4336	13.274	18.314
	6	.0050	.0101	.0254	.0517	5.1433	7.2598	10.925	14.544
	7	.0050	.0101	.0254	.0517	4.7374	6.5415	9.5466	12.404
	8	.0050	.0101	.0254	.0516	4.4590	6.0595	8.6491	11.042
	9	.0050	.0101	.0254	.0516	4.2565	5.7147	8.0215	10.107
	10	.0050	.0101	.0254	.0516	4.1028	5.4564	7.5594	9.4270
	12	.0050	.0101	.0254	.0515	3.8853	5.0959	6.9266	8.5096
	15	.0050	.0101	.0254	.0515	3.6823	4.7650	6.3589	7.7008
	20	.0050	.0101	.0253	.0514	3.4928	4.4613	5.8489	6.9865
	24	.0050	.0100	.0253	.0514	3.4028	4.3187	5.6136	6.6610
	30	.0050	.0100	.0253	.0514	3.3158	4.1821	5.3904	6.3547
	40	.0050	.0100	.0253	.0514	3.2317	4.0510	5.1785	6.0664
	60	.0050	.0100	.0253	.0513	3.1504	3.9253	4.9774	5.7950
	120	.0050	.0100	.0253	.0513	3.0718	3.8046	4.7865	5.5398
	∞	.0050	.0100	.0253	.0513	2.9957	3.6889	4.6052	5.2983

TABLE F.5   (*continued*)

NUM	DEN	.005	.01	.025	.05	.95	.975	.99	.995
DF					**PERCENTILE**				
3	1	.0180	.0293	.0573	.0987	215.71	864.16	5403.3	21615.
	2	.0201	.0324	.0623	.1047	19.164	39.165	99.166	199.17
	3	.0211	.0339	.0648	.1078	9.2766	15.439	29.457	47.467
	4	.0216	.0348	.0662	.1097	6.5914	9.9792	16.694	24.259
	5	.0220	.0354	.0672	.1109	5.4095	7.7636	12.060	16.530
	6	.0223	.0358	.0679	.1118	4.7571	6.5988	9.7795	12.917
	7	.0225	.0361	.0684	.1125	4.3468	5.8898	8.4513	10.882
	8	.0227	.0364	.0688	.1131	4.0662	5.4160	7.5910	9.5965
	9	.0228	.0366	.0691	.1135	3.8626	5.0781	6.9919	8.7171
	10	.0229	.0367	.0693	.1138	3.7083	4.8256	6.5523	8.0807
	12	.0230	.0370	.0697	.1144	3.4903	4.4742	5.9526	7.2258
	15	.0232	.0372	.0702	.1149	3.2874	4.1528	5.4170	6.4760
	20	.0234	.0375	.0706	.1155	3.0984	3.8587	4.9382	5.8177
	24	.0235	.0376	.0708	.1158	3.0088	3.7211	4.7181	5.5190
	30	.0235	.0377	.0710	.1161	2.9223	3.5894	4.5097	5.2388
	40	.0236	.0379	.0712	.1163	2.8387	3.4633	4.3126	4.9759
	60	.0237	.0380	.0715	.1167	2.7581	3.3425	4.1259	4.7290
	120	.0238	.0381	.0717	.1170	2.6802	3.2270	3.9493	4.4973
	∞	.0239	.0383	.0719	.1173	2.6049	3.1161	3.7816	4.2794
4	1	.0319	.0472	.0818	.1297	224.58	899.58	5624.6	22500.
	2	.0380	.0556	.0939	.1440	19.247	39.248	99.249	199.25
	3	.0412	.0599	.1002	.1517	9.1172	15.101	28.710	46.195
	4	.0432	.0626	.1041	.1565	6.3883	9.6045	15.977	23.155
	5	.0445	.0644	.1068	.1598	5.1922	7.3879	11.392	15.556
	6	.0455	.0658	.1087	.1623	4.5337	6.2272	9.1483	12.028
	7	.0462	.0668	.1102	.1641	4.1203	5.5226	7.8467	10.050
	8	.0468	.0676	.1114	.1655	3.8378	5.0526	7.0060	8.8051
	9	.0473	.0682	.1123	.1667	3.6331	4.7181	6.4221	7.9559
	10	.0477	.0687	.1131	.1677	3.4780	4.4683	5.9943	7.3428
	12	.0483	.0696	.1143	.1692	3.2592	4.1212	5.4119	6.5211
	15	.0489	.0704	.1155	.1707	3.0556	3.8043	4.8932	5.8029
	20	.0496	.0713	.1168	.1723	2.8661	3.5147	4.4307	5.1743
	24	.0499	.0718	.1175	.1732	2.7763	3.3794	4.2184	4.8898
	30	.0503	.0723	.1182	.1740	2.6896	3.2499	4.0179	4.6233
	40	.0506	.0727	.1189	.1749	2.6060	3.1261	3.8283	4.3738
	60	.0510	.0732	.1196	.1758	2.5252	3.0077	3.6491	4.1399
	120	.0514	.0738	.1203	.1767	2.4472	2.8943	3.4796	3.9207
	∞	.0517	.0743	.1211	.1777	2.3719	2.7858	3.3192	3.7151

TABLE F.5   *(continued)*

DF		PERCENTILE							
NUM	DEN	.005	.01	.025	.05	.95	.975	.99	.995
5	1	.0439	.0615	.0999	.1513	230.16	921.85	5763.7	23056.
	2	.0546	.0753	.1186	.1728	19.296	39.298	99.299	199.30
	3	.0605	.0829	.1288	.1849	9.0135	14.885	28.237	45.392
	4	.0643	.0878	.1354	.1926	6.2560	9.3645	15.522	22.456
	5	.0669	.0912	.1399	.1980	5.0503	7.1464	10.967	14.940
	6	.0689	.0937	.1433	.2020	4.3874	5.9876	8.7459	11.464
	7	.0704	.0956	.1459	.2051	3.9715	5.2852	7.4604	9.5221
	8	.0716	.0972	.1480	.2075	3.6875	4.8173	6.6318	8.3018
	9	.0726	.0984	.1497	.2095	3.4817	4.4844	6.0569	7.4711
	10	.0734	.0995	.1511	.2112	3.3258	4.2361	5.6363	6.8723
	12	.0747	.1011	.1533	.2138	3.1059	3.8911	5.0643	6.0711
	15	.0761	.1029	.1556	.2165	2.9013	3.5764	4.5556	5.3721
	20	.0775	.1047	.1580	.2194	2.7109	3.2891	4.1027	4.7616
	24	.0782	.1056	.1593	.2209	2.6207	3.1548	3.8951	4.4857
	30	.0790	.1066	.1606	.2224	2.5336	3.0265	3.6990	4.2276
	40	.0798	.1076	.1619	.2240	2.4495	2.9037	3.5138	3.9860
	60	.0806	.1087	.1633	.2257	2.3683	2.7863	3.3389	3.7600
	120	.0815	.1097	.1648	.2274	2.2900	2.6740	3.1735	3.5482
	∞	.0823	.1109	.1662	.2291	2.2141	2.5665	3.0173	3.3499
6	1	.0537	.0727	.1135	.1670	233.99	937.11	5859.0	23437.
	2	.0688	.0915	.1377	.1944	19.330	39.331	99.332	199.33
	3	.0774	.1022	.1515	.2102	8.9406	14.735	27.911	44.838
	4	.0831	.1093	.1606	.2206	6.1631	9.1973	15.207	21.975
	5	.0872	.1143	.1670	.2279	4.9503	6.9777	10.672	14.513
	6	.0903	.1181	.1718	.2334	4.2839	5.8197	8.4661	11.073
	7	.0927	.1211	.1756	.2377	3.8660	5.1186	7.1914	9.1554
	8	.0946	.1234	.1786	.2411	3.5806	4.6517	6.3707	7.9520
	9	.0962	.1254	.1810	.2440	3.3738	4.3197	5.8018	7.1338
	10	.0976	.1270	.1831	.2463	3.2172	4.0721	5.3858	6.5446
	12	.0997	.1296	.1863	.2500	2.9961	3.7283	4.8206	5.7570
	15	.1019	.1323	.1898	.2539	2.7905	3.4147	4.3183	5.0708
	20	.1043	.1352	.1935	.2581	2.5990	3.1283	3.8714	4.4721
	24	.1055	.1367	.1954	.2603	2.5082	2.9946	3.6667	4.2019
	30	.1069	.1383	.1974	.2626	2.4205	2.8667	3.4735	3.9492
	40	.1082	.1400	.1995	.2649	2.3359	2.7444	3.2910	3.7129
	60	.1096	.1417	.2017	.2674	2.2540	2.6274	3.1187	3.4918
	120	.1111	.1435	.2039	.2699	2.1750	2.5154	2.9559	3.2849
	∞	.1126	.1453	.2062	.2726	2.0986	2.4082	2.8020	3.0913

TABLE F.5 (*continued*)

DF					PERCENTILE				
NUM	DEN	.005	.01	.025	.05	.95	.975	.99	.995
7	1	.0616	.0817	.1239	.1788	236.77	948.22	5928.3	23715.
	2	.0806	.1047	.1529	.2111	19.353	39.355	99.356	199.36
	3	.0919	.1183	.1698	.2300	8.8868	14.624	27.672	44.434
	4	.0995	.1274	.1811	.2427	6.0942	9.0741	14.976	21.622
	5	.1050	.1340	.1892	.2518	4.8759	6.8531	10.456	14.200
	6	.1092	.1390	.1954	.2587	4.2066	5.6955	8.2600	10.786
	7	.1125	.1430	.2002	.2641	3.7870	4.9949	6.9928	8.8854
	8	.1152	.1462	.2041	.2684	3.5005	4.5286	6.1776	7.6942
	9	.1175	.1488	.2073	.2720	3.2927	4.1971	5.6129	6.8849
	10	.1193	.1511	.2100	.2750	3.1355	3.9498	5.2001	6.3025
	12	.1223	.1546	.2143	.2797	2.9134	3.6065	4.6395	5.5245
	15	.1255	.1584	.2189	.2848	2.7066	3.2934	4.1415	4.8473
	20	.1290	.1625	.2239	.2903	2.5140	3.0074	3.6987	4.2569
	24	.1308	.1646	.2265	.2932	2.4226	2.8738	3.4959	3.9905
	30	.1327	.1669	.2292	.2962	2.3343	2.7460	3.3045	3.7416
	40	.1347	.1692	.2321	.2994	2.2490	2.6238	3.1238	3.5088
	60	.1368	.1717	.2351	.3026	2.1665	2.5068	3.9530	3.2911
	120	.1390	.1743	.2382	.3060	2.0867	2.3948	3.7918	3.0874
	∞	.1413	.1770	.2414	.3096	2.0096	2.2875	3.6393	2.8968
8	1	.0681	.0888	.1321	.1880	238.88	956.66	5981.6	23925.
	2	.0906	.1156	.1650	.2243	19.371	39.373	99.374	199.37
	3	.1042	.1317	.1846	.2459	8.8452	14.540	27.489	44.126
	4	.1136	.1427	.1979	.2606	6.0410	8.9796	14.799	21.352
	5	.1205	.1508	.2076	.2712	4.8183	6.7572	10.289	13.961
	6	.1257	.1570	.2150	.2793	4.1468	5.5996	8.1016	10.566
	7	.1300	.1619	.2208	.2857	3.7257	4.8994	6.8401	8.6781
	8	.1334	.1659	.2256	.2909	3.4381	4.4332	6.0289	7.4960
	9	.1363	.1692	.2295	.2951	3.2296	4.1020	5.4671	6.6933
	10	.1387	.1720	.2328	.2988	3.0717	3.8549	5.0567	6.1159
	12	.1425	.1765	.2381	.3045	2.8486	3.5118	4.4994	5.3451
	15	.1467	.1813	.2438	.3107	2.6408	3.1987	4.0045	4.6743
	20	.1513	.1866	.2500	.3174	2.4471	2.9128	3.5644	4.0900
	24	.1538	.1894	.2533	.3210	2.3551	2.7791	3.3629	3.8264
	30	.1563	.1924	.2568	.3247	2.2662	2.6513	3.1726	3.5801
	40	.1590	.1955	.2604	.3286	2.1802	2.5289	2.9930	3.3498
	60	.1619	.1987	.2642	.3328	2.0970	2.4117	2.8233	3.1344
	120	.1649	.2022	.2682	.3370	2.0164	2.2994	2.6629	2.9330
	∞	.1680	.2058	.2725	.3416	1.9384	2.1918	2.5113	2.7444

TABLE F.5   (*continued*)

DF		PERCENTILE							
NUM	DEN	.005	.01	.025	.05	.55	.975	.99	.995
9	1	.0734	.0947	.1387	.1954	240.54	963.28	6022.5	24091.
	2	.0989	.1247	.1750	.2349	19.385	39.387	99.388	199.39
	3	.1147	.1430	.1969	.2589	8.8123	14.473	27.345	43.882
	4	.1257	.1557	.2119	.2752	5.9988	8.9047	14.659	21.139
	5	.1339	.1651	.2230	.2872	4.7725	6.6810	10.158	13.772
	6	.1402	.1724	.2315	.2964	4.0990	5.5234	7.9761	10.391
	7	.1453	.1782	.2383	.3037	3.6767	4.8232	6.7188	8.5138
	8	.1494	.1829	.2438	.3096	3.3881	4.3572	5.9106	7.3386
	9	.1529	.1869	.2484	.3146	3.1789	4.0260	5.3511	6.5411
	10	.1558	.1902	.2523	.3188	3.0204	3.7790	4.9424	5.9676
	12	.1606	.1956	.2585	.3254	2.7964	3.4358	4.3875	5.2021
	15	.1658	.2015	.2653	.3327	2.5876	3.1227	3.8948	4.5364
	20	.1715	.2080	.2727	.3405	2.3928	2.8365	3.4567	3.9564
	24	.1745	.2115	.2767	.3448	2.3002	2.7027	3.2560	3.6949
	30	.1778	.2151	.2809	.3492	2.2107	2.5746	3.0665	3.4505
	40	.1812	.2190	.2853	.3539	2.1240	2.4519	2.8876	3.2220
	60	.1848	.2231	.2899	.3588	2.0401	2.3344	2.7185	3.0083
	120	.1887	.2274	.2948	.3640	1.9588	2.2217	2.5586	2.8083
	∞	.1928	.2320	.3000	.3694	1.8799	2.1136	2.4073	2.6210
10	1	.0780	.0996	.1442	.2014	241.88	968.63	6055.8	24224.
	2	.1061	.1323	.1833	.2437	19.396	39.398	99.399	199.40
	3	.1237	.1526	.2072	.2697	8.7855	14.419	27.229	43.686
	4	.1362	.1668	.2238	.2875	5.9644	8.8439	14.546	20.967
	5	.1455	.1774	.2361	.3007	4.7351	6.6192	10.051	13.618
	6	.1528	.1857	.2456	.3108	4.0600	5.4613	7.8741	10.250
	7	.1587	.1923	.2532	.3189	3.6365	4.7611	6.6201	8.3803
	8	.1635	.1978	.2594	.3255	3.3472	4.2951	5.8143	7.2107
	9	.1676	.2023	.2646	.3311	3.1373	3.9639	5.2565	6.4171
	10	.1710	.2062	.2690	.3358	2.9782	3.7168	4.8492	5.8467
	12	.1766	.2125	.2762	.3433	2.7534	3.3736	4.2961	5.0855
	15	.1828	.2194	.2840	.3515	2.5437	3.0602	3.8049	4.4236
	20	.1896	.2270	.2925	.3605	2.3479	2.7737	3.3682	3.8470
	24	.1933	.2311	.2971	.3653	2.2547	2.6396	3.1681	3.5870
	30	.1972	.2355	.3020	.3704	2.1646	2.5112	2.9791	3.3440
	40	.2014	.2401	.3072	.3758	2.0772	2.3882	2.8005	3.1167
	60	.2058	.2450	.3127	.3815	1.9926	2.2702	2.6318	2.9042
	120	.2105	.2502	.3185	.3876	1.9105	2.1570	2.4721	2.7052
	∞	.2156	.2558	.3247	.3940	1.8307	2.0483	2.3209	2.5188

TABLE F.5 (continued)

DF | PERCENTILE

NUM	DEN	.005	.01	.025	.05	.95	.975	.99	.995
12	1	.0851	.1072	.1526	.2106	243.91	976.71	6106.3	24426.
	2	.1175	.1444	.1962	.2574	19.413	39.415	99.416	199.42
	3	.1384	.1680	.2235	.2865	8.7446	14.337	27.052	43.387
	4	.1533	.1848	.2426	.3068	5.9117	8.7512	14.374	20.705
	5	.1647	.1975	.2570	.3220	4.6777	6.5246	9.8883	13.384
	6	.1737	.2074	.2682	.3338	3.9999	5.3662	7.7183	10.034
	7	.1810	.2155	.2773	.3432	3.5747	4.6658	6.4691	8.1764
	8	.1871	.2222	.2847	.3510	3.2840	4.1997	5.6668	7.0149
	9	.1922	.2279	.2910	.3576	3.0729	3.8682	5.1114	6.2274
	10	.1966	.2328	.2964	.3632	2.9130	3.6209	4.7059	5.6613
	12	.2038	.2407	.3051	.3722	2.6866	3.2773	4.1553	4.9063
	15	.2118	.2494	.3147	.3821	2.4753	2.9633	3.6662	4.2498
	20	.2208	.2592	.3254	.3931	2.2776	2.6758	3.2311	3.6779
	24	.2257	.2645	.3313	.3991	2.1834	2.5412	3.0316	3.4199
	30	.2309	.2702	.3375	.4055	2.0921	2.4120	2.8431	3.1787
	40	.2365	.2763	.3441	.4122	2.0035	2.2882	2.6648	2.9531
	60	.2426	.2828	.3511	.4194	1.9174	2.1692	2.4961	2.7419
	120	.2491	.2899	.3588	.4272	1.8337	2.0548	2.3363	2.5439
	∞	.2561	.2976	.3670	.4355	1.7522	1.9447	2.1848	2.3583
15	1	.0926	.1152	.1613	.2201	245.95	984.87	6157.3	24630.
	2	.1299	.1573	.2099	.2716	19.429	39.431	99.432	199.43
	3	.1544	.1846	.2408	.3042	8.7029	14.253	26.872	43.085
	4	.1723	.2044	.2629	.3273	5.8578	8.6565	14.198	20.438
	5	.1861	.2195	.2796	.3447	4.6188	6.4277	9.7222	13.146
	6	.1972	.2316	.2928	.3584	3.9381	5.2687	7.5590	9.8140
	7	.2063	.2415	.3036	.3695	3.5108	4.5678	6.3143	7.9678
	8	.2139	.2497	.3126	.3787	3.2184	4.1012	5.5151	6.8143
	9	.2204	.2567	.3202	.3865	3.0061	3.7694	4.9621	6.0325
	10	.2261	.2628	.3268	.3931	2.8450	3.5217	4.5582	5.4707
	12	.2353	.2728	.3375	.4040	2.6169	3.1772	4.0096	4.7214
	15	.2457	.2839	.3494	.4161	2.4035	2.8621	3.5222	4.0698
	20	.2576	.2966	.3629	.4297	2.2033	2.5731	3.0880	3.5020
	24	.2641	.3036	.3703	.4371	2.1077	2.4374	2.8887	3.2456
	30	.2712	.3111	.3783	.4451	2.0148	2.3072	2.7002	3.0057
	40	.2789	.3193	.3868	.4537	1.9245	2.1819	2.5216	2.7811
	60	.2873	.3282	.3962	.4629	1.8364	2.0613	2.3523	2.5705
	120	.2965	.3379	.4063	.4730	1.7505	1.9450	2.1915	2.3727
	∞	.3067	.3486	.4175	.4841	1.6664	1.8326	2.0385	2.1868

TABLE F.5   (*continued*)

DF						PERCENTILE			
NUM	DEN	.005	.01	.025	.05	.95	.975	.99	.995
20	1	.1006	.1235	.1703	.2298	248.01	993.10	6208.7	24836.
	2	.1431	.1710	.2241	.2863	19.446	39.448	99.449	199.45
	3	.1719	.2025	.2591	.3227	8.6602	14.167	26.690	42.778
	4	.1933	.2257	.2845	.3489	5.8025	8.5599	14.020	20.167
	5	.2100	.2437	.3040	.3689	4.5581	6.3285	9.5527	12.903
	6	.2236	.2583	.3197	.3848	3.8742	5.1684	7.3958	9.5888
	7	.2349	.2704	.3325	.3978	3.4445	4.4667	6.1554	7.7540
	8	.2445	.2805	.3433	.4086	3.1503	3.9995	5.3591	6.6082
	9	.2528	.2893	.3525	.4179	2.9365	3.6669	4.8080	5.8318
	10	.2599	.2969	.3605	.4259	2.7740	3.4186	4.4054	5.2740
	12	.2719	.3095	.3737	.4391	2.5436	3.0728	3.8584	4.5299
	15	.2855	.3238	.3886	.4539	2.3275	2.7559	3.3719	3.8826
	20	.3014	.3404	.4058	.4708	2.1242	2.4645	2.9377	3.3178
	24	.3104	.3497	.4153	.4802	2.0267	2.3273	2.7380	3.0624
	30	.3202	.3599	.4258	.4904	1.9317	2.1952	2.5487	2.8230
	40	.3310	.3711	.4372	.5015	1.8389	2.0677	2.3689	2.5984
	60	.3430	.3835	.4498	.5138	1.7480	1.9445	2.1978	2.3872
	120	.3564	.3973	.4638	.5273	1.6587	1.8249	2.0346	2.1881
	∞	.3717	.4130	.4795	.5425	1.5705	1.7085	1.8783	1.9998
24	1	.1047	.1278	.1749	.2348	249.05	997.25	6234.6	24940.
	2	.1501	.1781	.2316	.2939	19.454	39.456	99.458	199.46
	3	.1812	.2119	.2687	.3324	8.6385	14.124	26.598	42.622
	4	.2045	.2371	.2959	.3602	5.7744	8.5109	13.929	20.030
	5	.2229	.2567	.3170	.3816	4.5272	6.2780	9.4665	12.780
	6	.2380	.2727	.3339	.3987	3.8415	5.1172	7.3127	9.4741
	7	.2506	.2860	.3480	.4128	3.4105	4.4150	6.0743	7.6450
	8	.2613	.2974	.3598	.4246	3.1152	3.9472	5.2793	6.5029
	9	.2706	.3071	.3700	.4347	2.9005	3.6142	4.7290	5.7292
	10	.2788	.3157	.3788	.4435	2.7372	3.3654	4.3269	5.1732
	12	.2924	.3299	.3935	.4580	2.5055	3.0187	3.7805	4.4315
	15	.3081	.3462	.4103	.4744	2.2878	2.7006	3.2940	3.7859
	20	.3265	.3652	.4297	.4934	2.0825	2.4076	2.8594	3.2220
	24	.3371	.3761	.4407	.5041	1.9838	2.2693	2.6591	2.9667
	30	.3487	.3880	.4527	.5157	1.8874	2.1359	2.4689	2.7272
	40	.3616	.4012	.4660	.5285	1.7929	2.0069	2.2880	2.5020
	60	.3761	.4161	.4808	.5428	1.7001	1.8817	2.1154	2.2898
	120	.3927	.4329	.4975	.5588	1.6084	1.7597	1.9500	2.0890
	∞	.4119	.4523	.5167	.5770	1.5173	1.6402	1.7908	1.8983

TABLE F.5 (*continued*)

DF						PERCENTILE			
NUM	DEN	.005	.01	.025	.05	.95	.975	.99	.995
30	1	.1089	.1322	.1796	.2398	250.09	1001.4	6260.7	25044.
	2	.1574	.1855	.2391	.3016	19.462	39.465	99.466	199.47
	3	.1909	.2217	.2786	.3422	8.6166	14.081	26.505	42.466
	4	.2163	.2489	.3077	.3718	5.7459	8.4613	13.838	19.892
	5	.2365	.2703	.3304	.3947	4.4957	6.2269	9.3793	12.656
	6	.2532	.2879	.3488	.4131	3.8082	5.0652	7.2285	9.3583
	7	.2673	.3026	.3642	.4284	3.3758	4.3624	5.9921	7.5345
	8	.2793	.3152	.3772	.4413	3.0794	3.8940	5.1981	6.3961
	9	.2898	.3261	.3884	.4523	2.8637	3.5604	4.6486	5.6248
	10	.2990	.3357	.3982	.4620	2.6996	3.3110	4.2469	5.0705
	12	.3146	.3517	.4146	.4780	2.4663	2.9633	3.7008	4.3309
	15	.3327	.3703	.4334	.4963	2.2468	2.6437	3.2141	3.6867
	20	.3542	.3924	.4555	.5177	2.0391	2.3486	2.7785	3.1234
	24	.3667	.4050	.4682	.5298	1.9390	2.2090	2.5773	2.8679
	30	.3805	.4191	.4822	.5432	1.8409	2.0739	2.3860	2.6278
	40	.3962	.4349	.4978	.5581	1.7444	1.9429	2.2034	2.4015
	60	.4141	.4529	.5155	.5748	1.6491	1.8152	2.0285	2.1874
	120	.4348	.4738	.5358	.5940	1.5543	1.6899	1.8600	1.9839
	∞	.4596	.4984	.5597	.6164	1.4591	1.5660	1.6964	1.7891
40	1	.1133	.1367	.1844	.2448	251.14	1005.6	6286.3	25148.
	2	.1648	.1931	.2468	.3094	19.471	39.473	99.474	199.47
	3	.2010	.2319	.2887	.3523	8.5944	14.037	26.411	42.308
	4	.2286	.2612	.3199	.3837	5.7170	8.4111	13.745	19.752
	5	.2509	.2846	.3444	.4082	4.4638	6.1751	9.2912	12.530
	6	.2693	.3039	.3644	.4281	3.7743	5.0125	7.1432	9.2408
	7	.2850	.3201	.3811	.4446	3.3404	4.3089	5.9084	7.4225
	8	.2985	.3341	.3954	.4587	3.0428	3.8398	5.1156	6.2875
	9	.3104	.3463	.4078	.4708	2.8259	3.5055	4.5667	5.5186
	10	.3208	.3571	.4187	.4814	2.6609	3.2554	4.1653	4.9659
	12	.3386	.3753	.4370	.4991	2.4259	2.9063	3.6192	4.2282
	15	.3596	.3966	.4583	.5196	2.2043	2.5850	3.1319	3.5850
	20	.3848	.4221	.4836	.5438	1.9938	2.2873	2.6947	3.0215
	24	.3997	.4371	.4983	.5578	1.8920	2.1460	2.4923	2.7654
	30	.4164	.4538	.5147	.5733	1.7918	2.0089	2.2992	2.5241
	40	.4356	.4730	.5333	.5907	1.6928	1.8752	2.1142	2.2958
	60	.4579	.4952	.5547	.6108	1.5943	1.7440	1.9360	2.0789
	120	.4846	.5216	.5800	.6343	1.4952	1.6141	1.7628	1.8709
	∞	.5177	.5541	.6108	.6627	1.3940	1.4835	1.5923	1.6691

TABLE F.5   (*continued*)

DF		.005	.01	.025	.05	.95	.975	.99	.995
NUM	DEN								
60	1	.1177	.1413	.1892	.2499	252.20	1009.8	6313.0	25253.
	2	.1726	.2009	.2548	.3174	19.479	39.481	99.483	199.48
	3	.2115	.2424	.2992	.3626	8.5720	13.992	26.316	42.149
	4	.2415	.2740	.3325	.3960	5.6878	8.3604	13.652	19.611
	5	.2660	.2995	.3589	.4222	4.4314	6.1225	9.2020	12.402
	6	.2864	.3206	.3806	.4437	3.7398	4.9589	7.0568	9.1219
	7	.3038	.2530	.3989	.4616	3.3043	4.2544	5.8236	7.3088
	8	.3190	.3542	.4147	.4769	3.0053	3.7844	5.0316	6.1772
	9	.3324	.3678	.4284	.4902	2.7872	3.4493	4.4831	5.4104
	10	.3443	.3800	.4405	.5019	2.6211	3.1984	4.0819	4.8592
	12	.3647	.4006	.4610	.5215	2.3842	2.8478	3.5355	4.1229
	15	.3890	.4251	.4851	.5445	2.1601	2.5242	3.0471	3.4803
	20	.4189	.4550	.5143	.5721	1.9464	2.2234	2.6077	2.9159
	24	.4367	.4727	.5314	.5882	1.8424	2.0799	2.4035	2.6585
	30	.4572	.4930	.5509	.6064	1.7396	1.9400	2.2079	2.4151
	40	.4810	.5165	.5734	.6272	1.6373	1.8028	2.0194	2.1838
	60	.5096	.5446	.6000	.6518	1.5343	1.6668	1.8363	1.9622
	120	.5452	.5793	.6325	.6815	1.4290	1.5299	1.6557	1.7469
	∞	.5922	.6248	.6747	.7198	1.3180	1.3883	1.4730	1.5325
120	1	.1223	.1460	.1941	.2551	253.25	1014.0	6339.4	25359.
	2	.1805	.2089	.2628	.3255	19.487	39.490	99.491	199.49
	3	.2224	.2532	.3099	.3731	8.5494	13.947	26.221	41.989
	4	.2551	.2874	.3455	.4086	5.6581	8.3092	13.558	19.468
	5	.2818	.3151	.3740	.4367	4.3984	6.0693	9.1118	12.274
	6	.3044	.3383	.3976	.4598	3.7047	4.9045	6.9690	9.0015
	7	.3239	.2637	.4176	.4792	3.2674	4.1989	5.7372	7.1933
	8	.3409	.3755	.4349	.4959	2.9669	3.7279	4.9460	6.0649
	9	.3561	.3908	.4501	.5105	2.7475	3.3918	4.3978	5.3001
	10	.3697	.4045	.4636	.5234	2.5801	3.1399	3.9965	4.7501
	12	.3931	.4280	.4867	.5454	2.3410	2.7874	3.4494	4.0149
	15	.4215	.4563	.5141	.5713	2.1141	2.4611	2.9595	3.3722
	20	.4570	.4915	.5480	.6029	1.8963	2.1562	2.5168	2.8058
	24	.4787	.5128	.5683	.6217	1.7897	2.0099	2.3099	2.5463
	30	.5041	.5376	.5918	.6434	1.6835	1.8664	2.1107	2.2997
	40	.5345	.5673	.6195	.6688	1.5766	1.7242	1.9172	2.0635
	60	.5724	.6040	.6536	.6998	1.4673	1.5810	1.7263	1.8341
	120	.6229	.6523	.6980	.7397	1.3519	1.4327	1.5330	1.6055
	∞	.6988	.7244	.7631	.7975	1.2214	1.2684	1.3246	1.3637

TABLE F.5   (continued)

| DF | | PERCENTILE | | | | | | | |
NUM	DEN	.005	.01	.025	.05	.95	.975	.99	.995
∞	1	.1269	.1507	.1990	.2603	254.32	1018.3	6366.0	25465.
	2	.1887	.2171	.2711	.3338	19.496	39.498	99.501	199.51
	3	.2337	.2644	.3209	.3839	8.5265	13.902	26.125	41.829
	4	.2692	.3013	.3590	.4216	5.6281	8.2573	13.463	19.325
	5	.2985	.3314	.3896	.4517	4.3650	6.0153	9.0204	12.144
	6	.3235	.3569	.4152	.4765	3.6688	4.8491	6.8801	8.8793
	7	.3452	.2748	.4372	.4976	3.2298	4.1423	5.6495	7.0760
	8	.3644	.3982	.4562	.5159	2.9276	3.6702	4.8588	5.9505
	9	.3815	.4154	.4731	.5319	2.7067	3.3329	4.3105	5.1875
	10	.3970	.4309	.4882	.5462	2.5379	3.0798	3.9090	4.6385
	12	.4240	.4577	.5142	.5707	2.2962	2.7249	3.3608	3.9039
	15	.4573	.4906	.5457	.6001	2.0658	2.3953	2.8684	3.2602
	20	.5001	.5324	.5853	.6367	1.8432	2.0853	2.4212	2.6904
	24	.5268	.5584	.6097	.6591	1.7331	1.9353	2.2107	2.4276
	30	.5589	.5895	.6386	.6854	1.6223	1.7867	2.0062	2.1760
	40	.5991	.6280	.6741	.7174	1.5089	1.6371	1.8047	1.9318
	60	.6525	.6789	.7203	.7587	1.3893	1.4822	1.6006	1.6885
	120	.7333	.7549	.7884	.8187	1.2539	1.3104	1.3805	1.4311
	∞	1.0000	1.0000	1.0000	1.0000	1.0000	1.0000	1.0000	1.0000

TABLE F.6   *.95 and .99 percentiles of F-max distributions*

K	DF	.95	.99	K	DF	.95	.99
3	2	87.5	448.	4	2	142.	729.
	3	27.8	85.		3	39.2	120.
	4	15.5	37.		4	20.6	49.
	5	10.8	22.		5	13.7	28.
	6	8.38	15.5		6	10.4	19.1
	7	6.94	12.1		7	8.44	14.5
	8	6.00	9.9		8	7.18	11.7
	9	5.34	8.5		9	6.31	9.9
	10	4.85	7.4		10	5.67	8.6
	12	4.16	6.1		12	4.79	6.9
	15	3.54	4.9		15	4.01	5.5
	20	2.95	3.8		20	3.29	4.3
	30	2.40	3.0		30	2.61	3.3
	60	1.85	2.2		60	1.96	2.3
	∞	1.00	1.0		∞	1.00	1.0
5	2	202.	1036.	6	2	266.	1362.
	3	50.7	151.		3	62.0	184.
	4	25.2	59.		4	29.5	69.
	5	16.3	33.		5	18.7	38.
	6	12.1	22.		6	13.7	25.
	7	9.70	16.5		7	10.8	18.4
	8	8.12	13.2		8	9.03	14.5
	9	7.11	11.1		9	7.80	12.1
	10	6.34	9.6		10	6.92	10.4
	12	5.30	7.6		12	5.72	8.2
	15	4.37	6.0		15	4.68	6.4
	20	3.54	4.6		20	3.76	4.9
	30	2.78	3.4		30	2.91	3.6
	60	2.04	2.4		60	2.11	2.4
	∞	1.00	1.0		∞	1.00	1.0

TABLE F.6   (continued)

K	DF	.95	.99	K	DF	.95	.99
7	2	333.	1705.	8	2	403.	2063.
	3	72.9	216.		3	83.5	249.
	4	33.6	79.		4	37.5	89.
	5	20.8	42.		5	22.9	46.
	6	15.0	27.		6	16.3	30.
	7	11.8	20.		7	12.7	22.
	8	9.78	15.8		8	10.5	16.9
	9	8.41	13.1		9	8.95	13.9
	10	7.42	11.1		10	7.87	11.8
	12	6.09	8.7		12	6.42	9.1
	15	4.95	6.7		15	5.19	7.1
	20	3.94	5.1		20	4.10	5.3
	30	3.02	3.7		30	3.12	3.8
	60	2.17	2.5		60	2.22	2.5
	∞	1.00	1.0		∞	1.00	1.0
9	2	475.	2432.	10	2	550.	2813.
	3	93.9	281.		3	104.	310.
	4	41.1	97.		4	44.6	106.
	5	24.7	50.		5	26.5	54.
	6	17.5	32.		6	18.6	34.
	7	13.5	23.		7	14.3	24.
	8	11.1	17.9		8	11.7	18.9
	9	9.45	14.7		9	9.91	15.3
	10	8.28	12.4		10	8.66	12.9
	12	6.72	9.5		12	7.00	9.9
	15	5.40	7.3		15	5.59	7.5
	20	4.24	5.5		20	4.37	5.6
	30	3.21	3.9		30	3.29	4.0
	60	2.26	2.6		60	2.30	2.6
	∞	1.00	1.0		∞	1.00	1.0

TABLE F.6   (*continued*)

K	DF	.95	.99	K	DF	.95	.99
11	2	626.	3204.	12	2	704.	3605.
	3	114.	337.		3	124.	361.
	4	48.0	113.		4	51.4	120.
	5	28.2	57.		5	29.9	60.
	6	19.7	36.		6	20.7	37.
	7	15.1	26.		7	15.8	27.
	8	12.2	19.8		8	12.7	21.
	9	10.3	16.0		9	10.7	16.6
	10	9.01	13.4		10	9.34	13.9
	12	7.25	10.2		12	7.48	10.6
	15	5.77	7.8		15	5.93	8.0
	20	4.49	5.8		20	4.59	5.9
	30	3.36	4.1		30	3.39	4.2
	60	2.33	2.7		60	2.36	2.7
	∞	1.00	1.0		∞	1.00	1.0

TABLE F.7   *Selected percentiles of distributions of product-moment correlation coefficients*

PERCENTILE

DF	.005	.01	.025	.05	.95	.975	.99	.995
1	-.9999	-.9995	-.9969	-.9877	.9877	.9969	.9995	.9999
2	-.9900	-.9800	-.9500	-.9000	.9000	.9500	.9800	.9900
3	-.9587	-.9343	-.8783	-.8053	.8053	.8783	.9343	.9587
4	-.9172	-.8822	-.8114	-.7293	.7293	.8114	.8822	.9172
5	-.8745	-.8329	-.7545	-.6694	.6694	.7545	.8329	.8745
6	-.8343	-.7888	-.7067	-.6215	.6215	.7067	.7888	.8343
7	-.7976	-.7498	-.6664	-.5823	.5823	.6664	.7498	.7976
8	-.7646	-.7154	-.6319	-.5495	.5495	.6319	.7154	.7646
9	-.7348	-.6850	-.6020	-.5214	.5214	.6020	.6850	.7348
10	-.7079	-.6581	-.5760	-.4972	.4972	.5760	.6581	.7079
11	-.6836	-.6339	-.5529	-.4762	.4762	.5529	.6339	.6836
12	-.6614	-.6120	-.5324	-.4574	.4574	.5324	.6120	.6614
13	-.6411	-.5922	-.5139	-.4409	.4409	.5139	.5922	.6411
14	-.6226	-.5742	-.4973	-.4258	.4258	.4973	.5742	.6226
15	-.6055	-.5577	-.4821	-.4124	.4124	.4821	.5577	.6055
16	-.5897	-.5425	-.4683	-.4000	.4000	.4683	.5425	.5897
17	-.5750	-.5285	-.4556	-.3888	.3888	.4556	.5285	.5750
18	-.5614	-.5154	-.4438	-.3783	.3783	.4438	.5154	.5614
19	-.5487	-.5033	-.4329	-.3687	.3687	.4329	.5033	.5487
20	-.5368	-.4921	-.4227	-.3599	.3599	.4227	.4921	.5368
21	-.5256	-.4816	-.4133	-.3516	.3516	.4133	.4816	.5256
22	-.5151	-.4715	-.4044	-.3438	.3438	.4044	.4715	.5151
23	-.5051	-.4623	-.3961	-.3365	.3365	.3961	.4623	.5051
24	-.4958	-.4534	-.3883	-.3297	.3297	.3883	.4534	.4958
25	-.4869	-.4451	-.3809	-.3233	.3233	.3809	.4451	.4869
26	-.4785	-.4372	-.3740	-.3173	.3173	.3740	.4372	.4785
27	-.4706	-.4297	-.3673	-.3114	.3114	.3673	.4297	.4706
28	-.4629	-.4226	-.3609	-.3060	.3060	.3609	.4226	.4629
29	-.4556	-.4158	-.3550	-.3009	.3009	.3550	.4158	.4556
30	-.4487	-.4093	-.3493	-.2959	.2959	.3493	.4093	.4487
40	-.3931	-.3578	-.3044	-.2573	.2573	.3044	.3578	.3931
50	-.3542	-.3218	-.2733	-.2306	.2306	.2733	.3218	.3542
60	-.3248	-.2948	-.2500	-.2109	.2109	.2500	.2948	.3248
80	-.2830	-.2565	-.2172	-.1829	.1829	.2172	.2565	.2830
100	-.2540	-.2302	-.1946	-.1638	.1638	.1946	.2302	.2540
120	-.2324	-.2104	-.1779	-.1496	.1496	.1779	.2104	.2324
200	-.1809	-.1636	-.1381	-.1161	.1161	.1381	.1636	.1809
500	-.1149	-.1038	-.0875	-.0735	.0735	.0875	.1038	.1149

TABLE F.8   *Transformation of correlation coefficients to z values*

	.000	.001	.002	.003	.004	.005	.006	.007	.008	.009
.00	0.0000	0.0010	0.0020	0.0030	0.0040	0.0050	0.0060	0.0070	0.0080	0.0090
.01	0.0100	0.0110	0.0120	0.0130	0.0140	0.0150	0.0160	0.0170	0.0180	0.0190
.02	0.0200	0.0210	0.0220	0.0230	0.0240	0.0250	0.0260	0.0270	0.0280	0.0290
.03	0.0300	0.0310	0.0320	0.0330	0.0340	0.0350	0.0360	0.0370	0.0380	0.0390
.04	0.0400	0.0410	0.0420	0.0430	0.0440	0.0450	0.0460	0.0470	0.0480	0.0490
.05	0.0500	0.0510	0.0520	0.0530	0.0541	0.0551	0.0561	0.0571	0.0581	0.0591
.06	0.0601	0.0611	0.0621	0.0631	0.0641	0.0651	0.0661	0.0671	0.0681	0.0691
.07	0.0701	0.0711	0.0721	0.0731	0.0741	0.0751	0.0761	0.0772	0.0782	0.0792
.08	0.0802	0.0812	0.0822	0.0832	0.0842	0.0852	0.0862	0.0872	0.0882	0.0892
.09	0.0902	0.0913	0.0923	0.0933	0.0943	0.0953	0.0963	0.0973	0.0983	0.0993
.10	0.1003	0.1013	0.1024	0.1034	0.1044	0.1054	0.1064	0.1074	0.1084	0.1094
.11	0.1104	0.1115	0.1125	0.1135	0.1145	0.1155	0.1165	0.1175	0.1186	0.1196
.12	0.1206	0.1216	0.1226	0.1236	0.1246	0.1257	0.1267	0.1277	0.1287	0.1297
.13	0.1307	0.1318	0.1328	0.1338	0.1348	0.1358	0.1368	0.1379	0.1389	0.1399
.14	0.1409	0.1419	0.1430	0.1440	0.1450	0.1460	0.1471	0.1481	0.1491	0.1501
.15	0.1511	0.1522	0.1532	0.1542	0.1552	0.1563	0.1573	0.1583	0.1593	0.1604
.16	0.1614	0.1624	0.1634	0.1645	0.1655	0.1665	0.1676	0.1686	0.1696	0.1706
.17	0.1717	0.1727	0.1737	0.1748	0.1758	0.1768	0.1779	0.1789	0.1799	0.1809
.18	0.1820	0.1830	0.1841	0.1851	0.1861	0.1872	0.1882	0.1892	0.1903	0.1913
.19	0.1923	0.1934	0.1944	0.1955	0.1965	0.1975	0.1986	0.1996	0.2006	0.2017
.20	0.2027	0.2038	0.2048	0.2059	0.2069	0.2079	0.2090	0.2100	0.2111	0.2121
.21	0.2132	0.2142	0.2153	0.2163	0.2174	0.2184	0.2195	0.2205	0.2216	0.2226
.22	0.2237	0.2247	0.2258	0.2268	0.2279	0.2289	0.2300	0.2310	0.2321	0.2331
.23	0.2342	0.2352	0.2363	0.2374	0.2384	0.2395	0.2405	0.2416	0.2427	0.2437
.24	0.2448	0.2458	0.2469	0.2480	0.2490	0.2501	0.2512	0.2522	0.2533	0.2543
.25	0.2554	0.2565	0.2575	0.2586	0.2597	0.2608	0.2618	0.2629	0.2640	0.2650
.26	0.2661	0.2672	0.2683	0.2693	0.2704	0.2715	0.2726	0.2736	0.2747	0.2758
.27	0.2769	0.2779	0.2790	0.2801	0.2812	0.2823	0.2833	0.2844	0.2855	0.2866
.28	0.2877	0.2888	0.2899	0.2909	0.2920	0.2931	0.2942	0.2953	0.2964	0.2975
.29	0.2986	0.2997	0.3008	0.3018	0.3029	0.3040	0.3051	0.3062	0.3073	0.3084
.30	0.3095	0.3106	0.3117	0.3128	0.3139	0.3150	0.3161	0.3172	0.3183	0.3194
.31	0.3205	0.3217	0.3228	0.3239	0.3250	0.3261	0.3272	0.3283	0.3294	0.3305
.32	0.3316	0.3328	0.3339	0.3350	0.3361	0.3372	0.3383	0.3395	0.3406	0.3417
.33	0.3428	0.3440	0.3451	0.3462	0.3473	0.3484	0.3496	0.3507	0.3518	0.3530
.34	0.3541	0.3552	0.3564	0.3575	0.3586	0.3598	0.3609	0.3620	0.3632	0.3643
.35	0.3654	0.3666	0.3677	0.3689	0.3700	0.3712	0.3723	0.3734	0.3746	0.3757
.36	0.3769	0.3780	0.3792	0.3803	0.3815	0.3826	0.3838	0.3850	0.3861	0.3873
.37	0.3884	0.3896	0.3907	0.3919	0.3931	0.3942	0.3954	0.3966	0.3977	0.3989
.38	0.4001	0.4012	0.4024	0.4036	0.4047	0.4059	0.4071	0.4083	0.4094	0.4106
.39	0.4118	0.4130	0.4142	0.4153	0.4165	0.4177	0.4189	0.4201	0.4213	0.4225

TABLE F.8 (*continued*)

	.000	.001	.002	.003	.004	.005	.006	.007	.008	.009
.40	0.4236	0.4248	0.4260	0.4272	0.4284	0.4296	0.4308	0.4320	0.4332	0.4344
.41	0.4356	0.4368	0.4380	0.4392	0.4404	0.4416	0.4428	0.4441	0.4453	0.4465
.42	0.4477	0.4489	0.4501	0.4513	0.4526	0.4538	0.4550	0.4562	0.4574	0.4587
.43	0.4599	0.4611	0.4624	0.4636	0.4648	0.4660	0.4673	0.4685	0.4698	0.4710
.44	0.4722	0.4735	0.4747	0.4760	0.4772	0.4784	0.4797	0.4809	0.4822	0.4834
.45	0.4847	0.4860	0.4872	0.4885	0.4897	0.4910	0.4922	0.4935	0.4948	0.4960
.46	0.4973	0.4986	0.4999	0.5011	0.5024	0.5037	0.5049	0.5062	0.5075	0.5088
.47	0.5101	0.5114	0.5126	0.5139	0.5152	0.5165	0.5178	0.5191	0.5204	0.5217
.48	0.5230	0.5243	0.5256	0.5269	0.5282	0.5295	0.5308	0.5321	0.5334	0.5347
.49	0.5361	0.5374	0.5387	0.5400	0.5413	0.5427	0.5440	0.5453	0.5466	0.5480
.50	0.5493	0.5506	0.5520	0.5533	0.5547	0.5560	0.5573	0.5587	0.5600	0.5614
.51	0.5627	0.5641	0.5654	0.5668	0.5681	0.5695	0.5709	0.5722	0.5736	0.5750
.52	0.5763	0.5777	0.5791	0.5805	0.5818	0.5832	0.5846	0.5860	0.5874	0.5888
.53	0.5901	0.5915	0.5929	0.5943	0.5957	0.5971	0.5985	0.5999	0.6013	0.6027
.54	0.6042	0.6056	0.6070	0.6084	0.6098	0.6112	0.6127	0.6141	0.6155	0.6169
.55	0.6184	0.6198	0.6213	0.6227	0.6241	0.6256	0.6270	0.6285	0.6299	0.6314
.56	0.6328	0.6343	0.6358	0.6372	0.6387	0.6401	0.6416	0.6431	0.6446	0.6460
.57	0.6475	0.6490	0.6505	0.6520	0.6535	0.6550	0.6565	0.6580	0.6595	0.6610
.58	0.6625	0.6640	0.6655	0.6670	0.6685	0.6700	0.6716	0.6731	0.6746	0.6761
.59	0.6777	0.6792	0.6807	0.6823	0.6838	0.6854	0.6869	0.6885	0.6900	0.6916
.60	0.6931	0.6947	0.6963	0.6978	0.6994	0.7010	0.7026	0.7042	0.7057	0.7073
.61	0.7089	0.7105	0.7121	0.7137	0.7153	0.7169	0.7185	0.7201	0.7218	0.7234
.62	0.7250	0.7266	0.7283	0.7299	0.7315	0.7332	0.7348	0.7365	0.7381	0.7398
.63	0.7414	0.7431	0.7447	0.7464	0.7481	0.7497	0.7514	0.7531	0.7548	0.7565
.64	0.7582	0.7599	0.7616	0.7633	0.7650	0.7667	0.7684	0.7701	0.7718	0.7736
.65	0.7753	0.7770	0.7788	0.7805	0.7823	0.7840	0.7858	0.7875	0.7893	0.7910
.66	0.7928	0.7946	0.7964	0.7981	0.7999	0.8017	0.8035	0.8053	0.8071	0.8089
.67	0.8107	0.8126	0.8144	0.8162	0.8180	0.8199	0.8217	0.8236	0.8254	0.8273
.68	0.8291	0.8310	0.8328	0.8347	0.8366	0.8385	0.8404	0.8422	0.8441	0.8460
.69	0.8480	0.8499	0.8518	0.8537	0.8556	0.8576	0.8595	0.8614	0.8634	0.8653
.70	0.8673	0.8693	0.8712	0.8732	0.8752	0.8772	0.8792	0.8812	0.8832	0.8852
.71	0.8872	0.8892	0.8912	0.8933	0.8953	0.8973	0.8994	0.9014	0.9035	0.9056
.72	0.9076	0.9097	0.9118	0.9139	0.9160	0.9181	0.9202	0.9223	0.9245	0.9266
.73	0.9287	0.9309	0.9330	0.9352	0.9373	0.9395	0.9417	0.9439	0.9461	0.9483
.74	0.9505	0.9527	0.9549	0.9571	0.9594	0.9616	0.9639	0.9661	0.9684	0.9707
.75	0.9730	0.9752	0.9775	0.9798	0.9822	0.9845	0.9868	0.9891	0.9915	0.9938
.76	0.9962	0.9986	1.0010	1.0034	1.0057	1.0082	1.0106	1.0130	1.0154	1.0179
.77	1.0203	1.0228	1.0253	1.0277	1.0302	1.0327	1.0352	1.0377	1.0403	1.0428
.78	1.0454	1.0479	1.0505	1.0531	1.0557	1.0583	1.0609	1.0635	1.0661	1.0688
.79	1.0714	1.0741	1.0768	1.0795	1.0822	1.0849	1.0876	1.0903	1.0931	1.0958

TABLE F.8   (*continued*)

	.000	.001	.002	.003	.004	.005	.006	.007	.008	.009
.80	1.0986	1.1014	1.1042	1.1070	1.1098	1.1127	1.1155	1.1184	1.1212	1.1241
.81	1.1270	1.1299	1.1329	1.1358	1.1388	1.1417	1.1447	1.1477	1.1507	1.1538
.82	1.1568	1.1599	1.1629	1.1660	1.1691	1.1723	1.1754	1.1786	1.1817	1.1849
.83	1.1881	1.1914	1.1946	1.1978	1.2011	1.2044	1.2077	1.2111	1.2144	1.2178
.84	1.2212	1.2246	1.2280	1.2314	1.2349	1.2384	1.2419	1.2454	1.2490	1.2526
.85	1.2561	1.2598	1.2634	1.2671	1.2707	1.2744	1.2782	1.2819	1.2857	1.2895
.86	1.2933	1.2972	1.3011	1.3050	1.3089	1.3129	1.3168	1.3209	1.3249	1.3290
.87	1.3331	1.3372	1.3414	1.3455	1.3498	1.3540	1.3583	1.3626	1.3670	1.3713
.88	1.3758	1.3802	1.3847	1.3892	1.3938	1.3984	1.4030	1.4077	1.4124	1.4171
.89	1.4219	1.4267	1.4316	1.4365	1.4415	1.4465	1.4515	1.4566	1.4618	1.4670
.90	1.4722	1.4775	1.4828	1.4882	1.4937	1.4992	1.5047	1.5103	1.5160	1.5217
.91	1.5275	1.5334	1.5393	1.5452	1.5513	1.5574	1.5636	1.5698	1.5761	1.5825
.92	1.5890	1.5956	1.6022	1.6089	1.6157	1.6226	1.6295	1.6366	1.6438	1.6510
.93	1.6584	1.6658	1.6734	1.6810	1.6888	1.6967	1.7047	1.7129	1.7211	1.7295
.94	1.7380	1.7467	1.7555	1.7644	1.7735	1.7828	1.7922	1.8018	1.8116	1.8216
.95	1.8317	1.8421	1.8527	1.8635	1.8745	1.8857	1.8972	1.9089	1.9210	1.9333
.96	1.9459	1.9588	1.9720	1.9856	1.9996	2.0139	2.0286	2.0438	2.0595	2.0756
.97	2.0922	2.1094	2.1272	2.1457	2.1648	2.1847	2.2053	2.2268	2.2493	2.2728
.98	2.2975	2.3234	2.3507	2.3795	2.4100	2.4426	2.4773	2.5146	2.5548	2.5986
.99	2.6465	2.6994	2.7585	2.8255	2.9028	2.9941	3.1059	3.2498	3.4525	3.7985

TABLE F.9   *10,000 random digits*

ROW
NUMBER

1	50691	91653	88574	08675	12700	32027	41034	56912	34264	77769
2	19787	66937	91769	13399	96096	43165	72096	86350	23062	99419
3	16746	77983	18061	23664	64557	78213	43857	68009	20483	00618
4	91039	16099	38824	00778	23058	76539	50584	71810	52589	32778
5	11075	62081	88977	78676	53855	56472	13090	01708	89016	45111
6	41230	92934	30342	29933	24597	72632	21727	63861	80454	47243
7	59028	24399	05075	64775	59803	45737	19025	46696	18914	03062
8	42957	25204	00753	60284	85482	34984	86637	95354	80698	87650
9	45881	59475	64445	97261	55252	50788	31295	16437	49497	22493
10	75104	45819	88471	75440	55309	63481	23616	64950	73291	10964
11	78614	07347	63528	84643	19455	95596	38158	75758	65628	10498
12	69279	59274	67459	53563	98241	18097	65297	49803	99145	25320
13	58626	91259	13832	75095	08333	53845	74223	82690	89320	89565
14	81630	00339	07996	65249	66792	05555	79169	12136	44621	95904
15	74330	13688	02044	65910	96007	82692	40473	56437	35671	95073
16	70829	66963	86390	26458	02385	41505	06239	68990	32915	89542
17	55084	58581	60759	20627	86682	76542	03648	28183	29823	68134
18	98845	17428	97397	62400	51284	92211	40593	82713	06067	46190
19	48116	91870	16346	97406	54649	42039	58407	84248	45780	60547
20	82778	31709	71564	26258	07522	03825	92087	21809	25678	39987
21	86615	67618	07446	63129	07111	70516	67289	09457	48995	08043
22	82558	99260	69136	35099	68187	85382	09569	94211	57824	98100
23	08290	70291	74080	96503	56140	27794	27765	51740	07712	29816
24	95062	76310	81603	86828	68370	46001	79205	35511	91239	52961
25	30361	66712	86801	29556	91232	98295	87322	99172	50009	27224
26	17390	96107	70391	78715	61943	33315	39778	97149	08122	86388
27	05390	33046	63920	28733	42644	38972	98161	79861	88282	28279
28	06624	21114	33869	20940	03732	39973	89948	81060	36381	06027
29	38146	77295	33742	00135	26587	54775	94846	18587	39327	71711
30	76430	23645	62335	60393	71813	52677	09917	89100	93855	75617
31	16664	30164	22546	63538	79376	26865	61996	60418	37777	84170
32	56424	64680	81038	79364	23815	44002	38480	09864	35960	10760
33	95954	15540	18554	63349	70259	03212	91950	16214	80378	56421
34	59007	56364	49965	61970	32493	55404	85950	99606	46328	17887
35	19341	87208	99853	40202	08553	78731	83463	19624	82512	13556
36	24505	87007	35748	54865	40209	49466	94574	31406	64422	87185
37	15086	92183	84632	36790	59608	00371	67456	55364	80669	75402
38	65664	02188	09164	70939	25856	24344	58859	10454	19212	59078
39	40397	76835	14062	96067	70645	23695	59140	75812	18804	55529
40	31700	24753	22919	43207	83387	27820	12494	30041	88927	22668

TABLE F.9   (*continued*)

ROW
NUMBER

41	14472	19372	23759	47116	81647	44946	97716	41157	30913	30842
42	18018	57089	98428	89075	77511	15194	69634	68269	52292	63404
43	16752	54266	76103	05268	41145	36100	73916	32462	01658	68565
44	47184	33660	96555	56656	18238	56888	29315	99813	47831	81385
45	93884	63945	06606	45545	29237	21040	43552	02749	19963	23705
46	44112	25139	62540	20200	38793	76373	01711	24722	31244	04592
47	10991	67616	96339	56810	45766	21363	57946	50002	07989	06215
48	34402	12087	57559	65862	61811	36453	08438	84435	75085	45182
49	61419	45244	32623	86912	50091	53116	59088	27149	21605	33372
50	84037	25936	04836	37968	76851	39660	10274	70731	45720	55762
51	09972	74036	23610	90926	40167	51840	28408	33628	08432	68823
52	33572	72847	33140	69914	24682	57531	45518	32677	92151	94317
53	30984	29000	17100	19781	66887	67105	69941	69372	19946	49800
54	89424	61867	19788	84639	53944	83243	82694	96001	64194	70157
55	24763	42978	60331	80383	54125	29033	06934	41093	20723	48368
56	07490	76138	66608	17033	38218	09644	71566	58802	05890	84586
57	82030	72988	67673	11222	92849	37809	96596	61390	01108	16682
58	97742	68757	19023	27385	17544	17864	54500	51633	03194	71313
59	79030	92963	47196	11086	30843	62492	54188	06620	03386	70633
60	38997	82566	49333	95115	13313	19834	21068	02220	85744	89839
61	46568	55793	11188	20315	77746	94728	56500	61830	52174	11940
62	93128	61298	66162	64056	62650	85380	99568	00405	69673	24628
63	35870	34391	38423	12661	27185	63510	16199	78509	90668	90576
64	70701	17920	27932	50483	35058	50046	74424	33870	18591	60743
65	96275	23095	03760	22726	86330	53401	25052	58613	54639	86291
66	72519	76643	42531	10541	37698	30425	57251	11422	82956	10225
67	74151	92320	23956	31195	63386	55061	64108	80458	92690	43296
68	75905	02954	96799	43216	38647	77622	92459	31591	66892	77107
69	43316	24900	89553	85381	17191	08877	97772	34217	67821	85869
70	16420	47818	47240	87248	95579	81827	12880	06675	95361	32638
71	06967	17690	57935	44906	81676	97422	76612	20982	99730	18202
72	48151	29322	95666	91640	06518	62750	25953	99952	78277	53471
73	76842	10952	17446	94373	82479	75033	29970	53456	75480	48236
74	24374	91591	90748	32626	48879	44069	09003	30063	33513	94415
75	41897	84117	12238	73620	30566	02551	20260	22856	41530	82461
76	24171	75567	24492	51938	99770	62409	73230	39369	76231	95334
77	99175	07282	69983	18034	48565	91887	68909	71676	61937	06058
78	16273	90115	31362	10790	39524	76553	96074	22961	67515	97496
79	57555	69188	94966	56747	29428	24038	31136	75291	76045	76800
80	97741	76023	94092	27731	26921	86064	13121	83138	99525	91197

TABLE F.9   (*continued*)

ROW
NUMBER

81	94337	22615	30349	42081	34739	24945	42200	06363	93259	77790
82	14038	70566	38431	19588	31832	65743	20009	84270	67370	44576
83	93639	69040	44900	62764	C9029	64724	91245	29387	91749	99954
84	63826	83075	38461	63C94	73666	63837	24549	89216	11686	21311
85	69178	91277	88645	83872	69880	23453	C7189	60745	37212	77893
86	66238	84321	89312	16843	7988C	23039	53334	44580	80370	59090
87	65142	19947	41183	95771	C3C05	33494	42078	65561	85287	95298
88	89932	98248	43809	30641	75089	40531	22015	01368	52197	28177
89	99C05	27907	53544	27451	38789	C9604	22999	15259	66875	01236
90	01271	98890	09743	33953	84817	94536	36219	39873	52202	61070
91	91594	30774	26558	87137	45299	4901C	92118	63330	83850	24098
92	49C57	074C9	24436	89740	36064	55875	14382	28297	76429	26706
93	75619	01003	40763	50803	87045	52475	25088	58912	48159	85432
94	91012	64270	26773	07419	65734	53608	14437	44524	35656	64196
95	25606	81035	19169	85C82	52592	80341	C0523	12387	21804	73385
96	52550	85022	01735	60130	14493	588C9	84679	55287	35106	35271
97	87784	37528	84134	08067	63558	957C6	66662	86033	59471	70393
98	01358	58995	32987	74145	23704	11259	22323	75820	62867	00985
99	13587	34452	22682	99577	72180	42534	19294	93614	83140	24863
100	65452	03054	62082	91152	29404	84458	88331	95510	93731	33865
101	29377	92654	16050	67662	79326	71902	24281	61582	42914	98515
102	29631	52576	84125	92393	37906	96505	37932	63447	51923	10435
103	79860	68088	66125	20523	13119	C0461	C2775	15220	64734	06424
104	85142	62453	22014	36547	57497	92186	39415	95108	72770	93320
105	53205	76546	87850	61C66	99040	56572	40891	63455	96259	89186
106	28315	11856	05562	58335	72507	6221U	68945	02576	30775	35309
107	86382	99013	C5265	30141	98751	37272	31315	68363	97597	14310
108	20663	86017	72482	73942	44089	50251	31965	09872	00559	69700
109	87809	24691	35104	08993	75548	95579	72987	90505	37636	31725
110	29148	36682	27676	65C61	94309	254C9	67453	40887	64668	98019
111	26466	20219	3816C	97C23	40495	15941	C4154	41248	84184	29969
112	70269	61982	40959	43190	90037	25552	64531	77203	99565	79820
113	78142	22127	76929	41275	51049	C4042	23557	96779	66090	80501
114	32C62	93330	01442	12382	49541	847C1	84220	47763	86011	58754
115	13448	16335	92331	28151	83667	738C4	C8C59	64630	36045	77082
116	04823	78538	91558	31821	77764	72765	52284	16749	25387	10313
117	16629	13553	43373	76924	16251	725C1	C4106	34691	15925	14447
118	75203	83865	1114C	52755	27047	1607C	56702	69870	67885	58133
119	75572	C4700	49960	44099	24914	93037	48409	34268	05635	71738
120	71810	45872	55319	65963	22225	72986	34204	08331	84245	86956

TABLE F.9   (*continued*)

ROW
NUMBER

121	71669	52192	79748	68287	42994	57813	17786	94999	68430	45318
122	73180	00869	64564	38830	75167	76855	07107	43893	88677	67682
123	94803	70892	03417	48408	92036	80869	47654	81124	31104	82485
124	32139	28146	91218	96167	93209	08028	91222	81373	23638	30262
125	48730	57557	43790	23723	88625	59158	44723	12268	49385	27529
126	91325	46822	82401	14504	71478	98772	76948	03481	33216	28454
127	09284	14508	98088	06714	95831	91461	45503	24084	42647	43729
128	07905	01338	34750	41426	63539	49613	91258	79146	17930	38689
129	89616	70054	75299	23685	54844	71640	48093	42916	27914	71947
130	78845	25762	13358	74285	24273	00111	01496	93323	00495	20261
131	69662	35846	00185	65650	57267	52365	60174	72634	98085	77689
132	27278	94450	46967	48816	36012	90068	94652	43683	65562	79692
133	73112	28587	88277	84056	47726	53198	82093	77509	69151	28983
134	05719	48476	35810	27207	84341	70599	34539	14381	10430	24404
135	51528	44241	44746	66132	69207	75724	60641	82077	76349	30308
136	84276	45150	44498	00220	60147	20131	75124	50023	23830	84629
137	77983	29368	38102	81516	46693	20586	80891	15408	96524	74525
138	49261	18008	55102	70516	10736	12187	67793	87139	85802	99319
139	63428	12397	66415	19953	68031	52395	19511	63914	82942	67452
140	89693	19532	94859	43709	94408	51477	54724	02567	71917	35816
141	70200	67205	46393	67355	65519	65268	86537	01757	79293	49148
142	97248	66980	60512	80216	43360	58252	65999	09810	44903	28592
143	41725	73141	32052	01991	30819	15304	91454	93285	54536	76825
144	97908	11006	07744	47451	15372	73359	64181	99072	25430	00383
145	24084	67802	82706	95797	32224	24759	96230	31699	64040	17136
146	49360	06562	36074	78792	23019	47047	24040	80223	33983	97628
147	10612	20943	90144	18510	95095	31411	64243	70708	46531	54521
148	78725	77546	42925	15538	77131	99647	61885	44499	47506	15546
149	04540	01150	23326	32149	41303	44862	03153	95341	95418	93674
150	10663	89501	26595	26796	42837	96609	93230	93394	08651	80618
151	65272	38499	96793	80589	51174	10156	58268	91636	23190	11101
152	78459	68233	13469	36132	80947	27031	30335	80142	24574	02155
153	72164	71687	45407	69032	57797	93757	84257	21725	11227	40230
154	26512	06875	26728	33340	21533	37842	80508	16015	33723	03570
155	81480	35368	29198	48232	01674	27272	39659	85224	99372	87308
156	38464	57010	83093	77480	28151	71807	76821	93602	78613	62428
157	97234	36368	33400	61207	82780	39089	35637	74898	69285	50315
158	17480	38257	49450	05855	44004	13895	27322	21617	75595	38624
159	09048	45024	07202	07299	69244	38103	53360	18922	87391	45133
160	66386	97952	50029	16896	35331	67880	38581	37969	78294	92987

TABLE F.9   (continued)

ROW
NUMBER

121	71669	52192	79748	68287	42994	57813	17786	94999	68430	45318
122	73180	00869	64564	38830	75167	76855	07107	43893	88677	67682
123	94803	70892	03417	48408	92036	80869	47654	81124	31104	82485
124	32139	28146	91218	96167	93209	08028	91222	81373	23638	30262
125	48730	57557	43790	23723	88625	59158	44723	12268	49385	27529
126	91325	46822	82401	14504	71478	98772	76948	03481	33216	28454
127	09284	14508	98088	06714	95831	91461	45503	24084	42647	43729
128	07905	01338	34750	41426	63539	49613	91258	79146	17930	38689
129	89616	70054	75299	23685	54844	71640	48093	42916	27914	71947
130	78845	25762	13358	74285	24273	00111	01496	93323	00495	20261
131	69662	35846	00185	65650	57267	52365	60174	72634	98085	77689
132	27278	94450	46967	48816	36012	90068	94652	43683	65562	79692
133	73112	28587	88277	84056	47726	53198	82093	77509	69151	28983
134	05719	48476	35810	27207	84341	70599	34539	14381	10430	24404
135	51528	44241	44746	66132	69207	75724	60641	82077	76349	30308
136	84276	45150	44498	00220	60147	20131	75124	50023	23830	84629
137	77983	29368	38102	81516	46693	20586	80891	15408	96524	74525
138	49261	18008	55102	70516	10736	12187	67793	87139	85802	99319
139	63428	12397	66415	19953	68031	52395	19511	63914	82942	67452
140	89693	19532	94859	43709	94408	51477	54724	02567	71917	35816
141	70200	67205	46393	67355	65519	65268	86537	01757	79293	49148
142	97248	66980	60512	80216	43360	58252	65999	09810	44903	28592
143	41725	73141	32052	01991	30819	15304	91454	93285	54536	76825
144	97908	11006	07744	47451	15372	73359	64181	99072	25430	00383
145	24084	67802	82706	95797	32224	24759	96230	31699	64040	17136
146	49360	06562	36074	78792	23019	47047	24040	80223	33983	97628
147	10612	20943	90144	18510	95095	31411	64243	70708	46531	54521
148	78725	77546	42925	15538	77131	99647	61885	44499	47506	15546
149	04540	01150	23326	32149	41303	44862	03153	95341	95418	93674
150	10663	89501	26595	26796	42837	96609	93230	93394	08651	80618
151	65272	38499	96793	80589	51174	10156	58268	91636	23190	11101
152	78459	68233	13469	36132	80947	27031	30335	80142	24574	02155
153	72164	71687	45407	69032	57797	93757	84257	21725	11227	40230
154	26512	06875	26728	33340	21533	37842	80508	16015	33723	03570
155	81480	35368	29198	48232	01674	27272	39659	85224	99372	87308
156	38464	57010	83093	77480	28151	71807	76821	93602	78613	62428
157	97234	36368	33400	61207	82780	39089	35637	74898	69285	50315
158	17480	38257	49450	05855	44004	13895	27322	21617	75595	38624
159	09048	45024	07202	07299	69244	38103	53360	18922	87391	45133
160	66386	97952	50029	16896	35331	67880	38581	37969	78294	92987

References

Baggaley, A. R.: "Mathematics of Introductory Statistics," Wiley, New York, 1969.

Boneau, C. A.: The Effects of Violations of Assumptions Underlying the *t* Test, *Psychol. Bull.*, **57**: 49–63 (1960).

Cochran, W. G.: The Comparison of Percentages in Matched Samples, *Biometrika*, **37**: 256–266 (1950).

————: The χ^2 Test of Goodness of Fit, *Ann. Math. Statist.*, **23**: 315–345 (1952).

————: Some Methods for Strengthening the Common χ^2 Tests, *Biometrics*, **10**: 417–451 (1954).

Dayton, C. M.: "An Empirical Evaluation of the Effects of Small Expected Values in 2 × 2 Contingency Tables on the Chi-square Test of Significance," Unpublished doctoral dissertation, University of Maryland, 1964.

————: "The Design of Educational Experiments," McGraw-Hill, New York, 1970.

Edwards, A. L.: On 'The Use and Misuse of the Chi-square Test'—The Case of the 2 × 2 Contingency Table, *Psychol. Bull.*, **47**: 341–346 (1950).

Fisher, R. A.: "Statistical Methods for Research Workers," Hafner, New York, 1946.

Knetz, W. J.: "An Empirical Study of the Effects of Selected Variables upon the Chi-square Distribution (AIR-R&D-1/63-FR)," American Institutes for Research, Washington, D.C., 1963.

Lewis, D., and Burke, C. J.: The Use and Misuse of the Chi-square Test, *Psychol Bull.,* **46:** 433–489 (1949).

————: Further Discussion of the Use and Misuse of the Chi-square Test, *Psychol. Bull.,* **47:** 347–355, (1950).

Lindquist, E. F.: "Design and Analysis of Experiments," Houghton Mifflin, Boston, 1953.

National Bureau of Standards: "Tables of the Binomial Probability Distribution," (Applied Mathematics Series 6), Government Printing Office, Washington, D.C., 1950.

Pastore, N.: Some Comments on 'The Use and Misuse of the Chi-square Test,' *Psychol. Bull.,* **47:** 338–340 (1950).

Peters, C. C.: The Misuse of Chi-square—A Reply to Lewis and Burke, *Psychol. Bull.,* **47:** 331–337 (1950).

Romig, H. G.: "50–100 Binomial Tables," Wiley, New York, 1953.

Siegel, S.: "Nonparametric Statistics for the Behavioral Sciences," McGraw-Hill, New York, 1956.

Stevens, S. S. (ed.): "Handbook of Experimental Psychology," Wiley, New York, 1951.

Walker, H. M.: "Mathematics Essential for Elementary Statistics," Holt, New York, 1951.

Solutions to selected problems

Chapter 1

1.2.   a. Continuous   d. Continuous
       b. Continuous   e. Discrete
       c. Discrete   f. Continuous

1.3.   Answer c is continuous, since distance is infinitely divisible.

1.4.   Answer b.

1.5.   Answer c.

1.6.   a. 27.2   c. 6.5
       b. 45.6   d. 24.5

1.7.   a. 24.
       b. 11.
       c. 14.

1.8.   a. Factorial sign   d. Equal
       b. Greater than   e. Unequal
       c. Less than   f. Equal to or greater than

g. Equal to or less than
h. Absolute value
i. Summation operator

j. ith value of a variable
k. Multiplication operator

1.9. a. $1 + 2 + 3 + 4 + 5 + 6 + 7 + 8 = 36$

b. $Y_2 + Y_3 + Y_4$

d. $d \sum_{i=1}^{4} X_i + 4c$

c. $X_1Y_1 + X_2Y_2 + X_3Y_3$

e. $5 \cdot 4 \cdot 3 \cdot 2 \cdot 1 = 120$

1.10. Measures are from an entire population (description) or from a random sample (inference); number of measured variables (univariate, bivariate, multivariate); number of groups of subjects (one, or two or more); scale of measurement of the quantified variables (nominal, ordinal, interval, ratio)

Chapter 2

2.1. $\frac{7}{10}$ or .7

2.2. a. $2(\frac{1}{3})(\frac{2}{3}) = \frac{4}{9}$

b. $2(\frac{10}{30})(\frac{20}{29}) = \frac{40}{87}$

2.3. $1/280$

2.4. 120

2.5. $P(X = 3) = [(8!)/(3!5!)](\frac{1}{2})^3(\frac{1}{2})^5 = 56(\frac{1}{256}) = \frac{7}{32}$, or .219

2.8. Bar diagram and pie diagram

2.9. The sample must be drawn from the population in such a way that each member of the population has an equal and independent chance of being included in the sample. It is the process of selection which determines whether or not a sample is random.

2.13. (a) Data must be represented in two or more mutually exclusive and exhaustive categories; and (b) some method for determining expected frequencies must be available.

2.14. Small; large; large; small

2.15. a. Chi-square $= 3.5^2/5 + -3.5^2/5 = 24.50/5 = 4.90$

b. Chi-square $= 4^2/5 + -4^2/5 = \frac{32}{5} = 6.40$

2.16. a. Chi-square $= 14.5^2/100 + -14.5^2/100 = 420.50/100 = 4.20$; this value also exceeds the tabular critical value and H_0 is rejected.

b. Chi-square $= 15^2/100 + -15^2/100 = \frac{450}{100} = 4.50$; with 1 degree of freedom, the tabular critical value at the .05 level is 3.84; thus H_0 is rejected.

2.17. Chi-square $= 2.39$, which is nonsignificant with 4 degrees of freedom.

Chapter 3

3.1. Total number of females is $150 - 50 = 100$; total number not getting A's is $150 - 30 = 120$; since $\frac{100}{150} = \frac{2}{3}$ of the total group is females, we would expect $\frac{2}{3}$ of those not receiving A's to be females; this is $(\frac{2}{3})(120) = 80$.

3.2.   Phi = .2357

3.3.   Chi-square = 4.50; with 1 degree of freedom, this value is significant at the .05 level.

3.4.   Chi-square = .388; with 2 degrees of freedom, this value is nonsignificant.

3.5.   If r is the number of rows, and c is the number of columns, the degrees of freedom are $(r - 1)(c - 1)$.

3.6.   Chi-square = 1.492; with 1 degree of freedom, this value is non-significant.

3.7.   Chi-square = 6.04; with 1 degree of freedom, this value is significant at the .05 level.

3.8.   Chi-square = 5.625; with 2 degrees of freedom, this value is non-significant.

Chapter 4

4.3. $P_{60} = 37.75$; $P_{80} = 41.50$

4.4.   From raw scores: mean = 36.34; median = 36.50; mode = 37.   From frequency distribution: mean = 36.34; median = 36.50; mode = 37.

4.5.   From raw scores: range = 28; $Q = 4.21$; variance = 38.82; standard deviation = 6.23.   From frequency distribution: range = $50.5 - 20.5 = 30.0$; $Q = 4.09$; variance = 37.54; standard deviation = 6.12.

4.6.   Skewness index = $3(\mu - P_{50})/\sigma = 3(36.34 - 36.50)/6.23 = -.48/6.23 = -.08$.

4.7.   Mean = 67.00 inches; variance = 18.00; standard deviation = 4.24 inches.

4.8.   Mean = 18.00 inches; variance = 36.00; standard deviation = 6.00 inches.

4.10.   Mean = 17 inches; variance = 18.00; standard deviation = 4.24.

4.11.   Mean = 70; variance = 200.

4.12.   Mean = 74.93; variance = 83.93.

Chapter 5

5.1.   a. $P(.75 < z < 1.68) = .1801$, b. $P(z > -.26) = .6026$.

5.2.   a. .0228, b. .6826.

5.3.   .9463

5.4. $\chi^2 = 19(100)/225 = 8.44$; with 19 degrees of freedom, the computed value is less than the tabular value at the 5th percentile and we conclude that the sample and population variances are not consistent.

5.5.   Unbiasedness, consistency, and efficiency.

5.6. Mean of 100 and standard error of $20\!\!/\!\!5 = 4$.

5.7. $z = -1.26$; class does not differ significantly from the county norm.

5.8. $t = 2.67$; with 24 degrees of freedom, reject the null hypothesis at the .05 level but accept at the .01 level (nondirectional test).

5.9. $t = 5.51$; with 16 degrees of freedom, reject the null hypothesis at the .01 level (nondirectional test).

Chapter 6

6.2.   a. $F_{4,9,.005} = .0473$; $F_{4,9,.995} = 7.9559$.

     b. $F_{3,5,.005} = .0220$; $F_{3,5,.995} = 16.530$.

     c. $F_{8,8,.005} = .1334$; $F_{8,8,.995} = 7.4960$.

6.3.   a. $F_{2,27,.95} = 3.3593$ (by interpolation).

     b. $F_{4,95,.95} = 2.4797$ (by interpolation).

     c. $F_{5,6,.95} = 4.3874$.

6.4.   a. $S_1{}^2 = 50.00$; $S_2{}^2 = 48.80$.

     b. $F = 1.025$ with 9 and 15 degrees of freedom; accept the hypothesis of homogeneity of variance.

6.5. $S^2_{max} = 17.50$; $S^2_{min} = 5.25$; $F_{max} = 3.33$; this is nonsignificant for both 6 groups and 3 degrees of freedom per group and 6 groups and 14 degrees of freedom per group.

6.6. $S_1{}^2 = 9.47$; $S_2{}^2 = 6.80$; $F = 1.39$, which, with 15 and 15 degrees of freedom, is nonsignificant at the .10 level.

6.8.   a. Three treatment levels

     b. Six subjects per group

     c. $\mu_1 = \mu_2 = \mu_3$

     d. $F = 48.71/12.37 = 3.94$ which is significant at the .05 level but not at the .01 level of significance.

6.9. $MS_{treat} = 46.30$; $MS_{error} = 12.15$; $F = 3.81$, which, with 2 and 27 degrees of freedom, is significant at the .05 level but not at the .01 level of significance.

6.10. $MS_{treat} = 67.82$; $MS_{error} = 61.71$; $F = 1.10$, which, with 3 and 36 degrees of freedom, is nonsignificant at the .05 level of significance.

Chapter 7

7.2.   b. $\hat{Y}_i = -35.60 + .92X_i$

     c. 95.04

     e. $r = .95$ which is significantly different from zero.

     f. $Sy.x = 4.48$

7.3.   a. $\hat{Y}_i = 17.60 + .96X_i$

     b. 43.52

7.4. 16 percent of the variance in Y is accounted for and 84 percent remains unaccounted for. The standard error of estimate is 8 percent less than the standard deviation of Y.

7.5. a. $\hat{Y}_i = 1.52 + 1.08X_i$

   b. $\hat{Y}_i = -.32 + .79X_i$

7.6. $\hat{Y}_i = 79.39 + .29X_i$

7.7. $t = 5.71$ with 58 degrees of freedom; this is significant at the .01 level with a nondirectional test.

7.8. $z = -3.06$, which is significant at the .01 level with a nondirectional test.

7.9. $z = -2.07$, which is significant at the .05 level.

Chapter 8

8.4. Using the Kolmogorov-Smirnov one-sample test, the maximum difference between an observed and a theoretical cumulative proportion is .27, which occurs for the low category. For a sample of size 35, the critical value at the .05 level of significance is .23, thus, we reject the null hypothesis.

8.5. Rho $= .74$; for a sample of size 10, the tabular critical value is .632 at the .05 level of significance and we conclude that the observed correlation is significantly different from 0.

8.9. Convert each set of scores to proportions (or percentages) and compare the groups.

8.10. The sum of ranks for group 1 is 37.5 and for group 2 is 40.5; for unequal group sizes, divide each sum by the appropriate group size.

8.11. The maximum difference in cumulative proportion between the two groups is .30 and this occurs for the C classification. Since this value does not exceed the critical value of .43, we conclude that there is no evidence that the ages are not distributed the same.

8.12. $u_1 = 19.5$; $u_2 = 16.5$; the computed z value is .24 which is nonsignificant.